PEST CONTROL:
An Assessment of
Present and Alternative
Technologies

VOLUME III

Cotton Pest Control

The Report of the Cotton Study Team
Study on Problems of Pest Control
Environmental Studies Board
National Research Council

NATIONAL ACADEMY OF SCIENCES
Washington, D.C. 1975

This is Volume III of *Pest Control: An Assessment of Present and Alternative Technologies*. The other volumes are:

Volume I *Contemporary Pest Control Practices and Prospects: The Report of the Executive Committee*
Volume II *Corn/Soybeans Pest Control*
Volume IV *Forest Pest Control*
Volume V *Pest Control and Public Health*

NOTICE: The project that is the subject of this report was approved by the Governing Board of the National Research Council, whose members are drawn from the Councils of the National Academy of Sciences, the National Academy of Engineering, and the Institute of Medicine. The members of the Committee responsible for the report were chosen for their special competences and with regard for appropriate balance.

This report has been reviewed by a group other than the authors according to procedures approved by a Report Review Committee consisting of members of the National Academy of Sciences, the National Academy of Engineering, and the Institute of Medicine.

Library of Congress Catalog Card No. 75-37180

International Standard Book No. 0-309-02412-9 (this volume)
0-309-02409-9 (entire set)

Available from

Printing and Publishing Office
National Academy of Sciences
2101 Constitution Avenue, N.W.
Washington, D. C. 20418

Printed in the United States of America

PREFACE

This study originated in discussions in the Executive Offices of the National Academy of Sciences in March 1971. The Environmental Studies Board (ESB)* subsequently continued the discussions and appointed a Planning Committee to develop a detailed proposal. The Planning Committee, sponsored by funds from the Scaife Family Charitable Trusts, submitted its report to the ESB in April 1972. The ESB approved the report and appointed Dr. Donald Kennedy of Stanford University as Chairman of the Executive Committee of the Study on Problems of Pest Control: A Technology Assessment. Financial support for the study was secured from the U.S. Department of Agriculture, the U.S. Environmental Protection Agency, and the Ford Foundation. We thank these public and private agencies for their assistance.

The Executive Committee was charged with assessing the diverse effects on society that arise from efforts to

---

*The National Academy of Sciences complex (National Academy of Sciences, National Academy of Engineering, Institute of Medicine) functions as official but independent advisor to the federal government in matters relating to science and technology through its operating agency, the National Research Council. The National Research Council comprises four Assemblies and four Commissions each of which deals with specific disciplines or societal problems. The Commission on Natural Resources, formed in 1973, oversees the work of five boards, among them the Environmental Studies Board. Thus the pest control study, having originated as described above, completed its activities under the aegis of the Commission on Natural Resources.

control pests and with evaluating long-range alternatives available for accomplishing pest control. The charge also called explicitly for an examination of social and institutional factors important to the conduct of pest control practices, factors that must be considered in any recommendations for changes in pest control policy. The Planning Committee and the Executive Committee concluded that the most effective way to examine pest control problems would be through a series of case studies. Four such studies were selected. Three of them centered on specific agroecosystems: cotton, corn and soybeans, and forests. The fourth case study dealt with issues related to the control of pests of public health significance in the United States and abroad. Each case study examined the pest species involved, current control measures, and the benefits and costs of controlling the pests by various techniques.

This volume presents the report of the Cotton Study Team. The reports of the Executive Committee and the other study teams are all available from the National Academy of Sciences in five volumes under the general title *Pest Control: An Assessment of Present and Alternative Technologies*. Volume I comprises the report of the Executive Committee and all the summaries and recommendations of the four study team reports. The report of the Executive Committee, in addition to drawing on the findings of the individual study teams, covers other areas not specifically treated by those teams. The remaining four volumes are bound separately as follows: Volume II: *Corn/Soybeans Pest Control;* Volume III: *Cotton Pest Control;* Volume IV: *Forest Pest Control;* and Volume V: *Pest Control and Public Health.*

ACKNOWLEDGMENTS

Howard Engstrom, of the National Research Council staff, attended all meetings of the study team, assisted in arrangements, and participated in the preparation of the final report. Dr. John Perkins served as principal staff officer for the study as a whole, and Bette Lou Fields took over his responsibilities in the final phases of the study. Linda Jones, Lucette Comer, Mary Dong, Susan Lesser, Christina Neumann, and Estelle Miller provided able and tireless secretarial support. We are grateful for their generous and effective contributions.

COTTON STUDY TEAM

Stanley D. Beck, *Chairman*
W. A. Henry Professor
Department of Entomology
University of Wisconsin
Madison, Wisconsin

Perry Adkisson
Professor and Head
Department of Entomology
Texas A&M University
College Station, Texas

Donald C. Erwin
Professor of Plant Pathology
University of California
Riverside, California

Lewis W. Jones
Professor of Sociology
Director, Research Manpower Training and Research Project
Tuskegee Institute
Alabama

William W. Murdoch
Professor of Biology
University of California
Santa Barbara, California

Richard B. Norgaard
Assistant Professor of Agricultural Economics
University of California
Berkeley, California

James S. Plaxico
Professor and Head
Department of Agricultural Economics
Oklahoma State University
Stillwater, Oklahoma

Paul W. Santelmann
Regant's Professor
Agronomy Department
Oklahoma State University
Stillwater, Oklahoma

Christine A. Choemaker
Assistant Professor of Environmental Engineering
Cornell University
Ithaca, New York

A. Dan Tarlock
Professor of Law
Indiana University School of Law
Bloomington, Indiana

Bradford Waddle
Distinguished Professor of Agronomy
University of Arkansas
Fayetteville, Arkansas

CONTENTS

SUMMARY, CONCLUSIONS, AND RECOMMENDATIONS

In the decade following World War II, modern synthetic pesticides and other technological developments had a great impact on cotton yields. By reducing the risk of disastrous crop losses from pests, pesticides encouraged the adoption of other highly productive practices such as fertilization, irrigation and drainage, and planting of long-growing indeterminate varieties of cotton. Profitable cotton production in the United States now depends upon the successful application of good management practices to a wide range of soil and climatic conditions. Throughout the Cotton Belt, efficient pest management practices have been and will continue to be of major importance.

The range of pest control technologies in practice today has developed, of course, in response to the changing insect, weed, and disease problems that have plagued the cotton industry. However, economic and social forces have had a considerable impact on the adoption of pest management and other cotton production practices, including government programs supporting farm incomes, registering the use of pesticides, and enforcing worker health, safety, and other standards. They also include the experience and education of farmers and their access to information on changing farming practices. These forces influence regional production patterns and provide incentives for the adoption of alternative production practices, both of which affect the relative importance of the key pest problems in the major producing regions. Too often, the indirect effects of social and economic forces on pest control problems and practices have gone unrecognized by those responsible for formulating agricultural policies directing research on pest management.

A wide range of measures are used to control cotton diseases, but the most effective method of control is the incorporation of genetic resistance into commercial varieties. Except for treatment of cottonseed for seedling diseases, the application of currently available chemicals is often uneconomical, and the total acreage treated with fungicides and fumigants is relatively small. Genetic resistance to root knot nematodes has been incorporated in some varieties, but soil fumigation is more effective against nematodes.

Two potential alternatives to control of cotton diseases are planting at optimum soil temperatures for rapid emergence and development of biological control methods. Late planting, while providing disease control, often conflicts with other desirable production and pest control practices, especially insect control.

Herbicides and tillage practices are the most common weed control method. Preplanting tillage (seedbed preparation) removes germinated seedlings but does not deplete seed reserves. Postplanting tillage (cultivation) removes weeds but may prune roots and injure plants. Herbicides can be applied before planting, before emergence, or after emergence, and their use has significantly reduced the hand labor required in cotton production. The weed spectrum is changing, partly because of the widespread use of selective herbicides. In many cases, grasses are declining in importance, while some broadleaf species in the same taxonomic family as cotton are becoming more prevalent. In addition, the encroachment of perennial weed species has become more serious. To overcome these problems, the practice of rotating herbicides is gaining acceptance.

The potential alternatives to herbicides and tillage in cotton weed control involve some form of biological or other mechanical control. Development of these techniques does not appear to have reached the stage where their advantages or liabilities as practical alternatives can be evaluated. Future weed control in cotton will likely develop toward a concept of total farm vegetation management. This would include actions to reduce weeds and weed seed populations in other crops, set-aside acres, and fence rows. Crop rotations and herbicide rotations might be an integral part of future weed control programs.

In every cotton-producing region of the United States, there are one or more major insect pests that, under our present state of knowledge, usually must be controlled with insecticides if cotton is to be grown profitably. Contemporary cotton production utilizes more insecticides than any other single crop. However, the use of these insecticides can create problems more serious than the damage caused by the major pests, such as target pest resurgence, secondary pest outbreaks, and pest resistance to insecticides. In fact, most of the important arthropod pests of cotton have developed resistance to one or more insecticides.

The alternatives to continued heavy reliance on insecticides include cultural practices, biological control, and genetic manipulation techniques in combination with reduced quantities of insecticides and other chemicals. These combinations of alternative practices are designed to manipulate the environment in order to make it as unfavorable as possible for the pest species (or more favorable for the entomophagous species); to reduce the rate of pest increase and damage; or to concentrate pest numbers in small areas where direct control measures may be applied with minimum disruption to the entomophagous species.

Cultural practices, combined with chemical control, have been developed into successful programs for reducing overwintering populations of two major cotton pests, the boll weevil and the pink bollworm. These programs involve shorter growing periods and earlier harvest and disposal of crop residue. Although the application of insecticides, defoliants, and desiccants is still required, the amounts needed are greatly reduced. There is no apparent barrier to wide-scale acceptance by cotton producers, and the programs should be encouraged.

Trap cropping, with or without pheromone, can be used in cultural insect control. The weevils are then killed in the trap crop with a limited amount of insecticides. This practice shows great promise as a control measure, especially when it complements the cultural practices directed at overwintering weevils. However, the practical application of these techniques has not been developed to the point that they can be recommended to farmers.

Varietal modification of the cotton plant offers additional promise for pest suppression. Early-maturing, rapidly fruiting cotton varieties may be able to produce a crop before the population of boll weevils rises to damaging levels. In addition, they may mature so fast that harvesting and disposal of crop residues are completed before environmental conditions force boll weevils and pink bollworms into diapause.

A frego-bract cotton variety has been developed that is resistant to the boll weevil, but it is still susceptible to attack by the cotton fleahopper and *Lygus* spp. A nectarless variety moderately resistant to lygus and fleahoppers and slightly resistant to *Heliothis* spp. has been developed. Nectarless and frego-bract characters may be combined in a stock that would be resistant to the boll weevil, fleahopper, lygus, and heliothis. Resistant and short-season varieties will have their greatest utility in integrated control programs that use cultural, chemical, and biological pest suppression techniques.

The immense value of natural biological control was not fully appreciated by entomologists or cotton producers until after the introduction of synthetic organic insecticides. When the effectiveness of the approximately 600 species of predators and parasites found in cotton is disrupted, often by a single insecticide application, many secondary pests do considerable damage to the crop. Attempts at using insect natural enemies have taken the form of importation of exotic species and augmentation of native species. Importations have not been successful in controlling any major cotton insects or weeds. The programmed release of laboratory-reared insects has been successful only in experimental plots. Techniques for the large-scale rearing and release of insect natural enemies have not yet been developed. The use of pathogens, such as *Bacillus thuringiensis* and the nuclear polyhedrosis viruses, have a great potential; however, more research is needed to produce a more reliable and consistent level of control and to improve the safety, production, formulation, and application of these microbial insecticides.

Recently, sophisticated combinations of cultural practices, biological controls, genetic manipulations, and chemical control techniques have been developed into proposed programs whose objective has been the eradication of specific cotton pests. The Agricultural Act of 1973 directs the Secretary of Agriculture to eradicate the boll weevil and other insects, if feasible.

We recommend that the development of these alternative pest control practices into effective cotton pest management programs be encouraged. Although a comprehensive plan for eradicating the boll weevil from presently infested areas and to institute a containment program to prevent reentry from Mexico has been developed by USDA and state and industry scientists concerned with the boll weevil problem, the Cotton Study Team seriously questions whether technology is currently available to achieve this goal. Any consideration of such a difficult and costly program should await a critical analysis of the results of a proposed prior trial eradication program in North Carolina.

One feature of programs aimed at eradication is that the costs may be high and exceed benefits during the period that the pest is pushed toward extinction. When very low population densities are reached, benefits may be substantial. When the eradication program is relaxed, however, undetected insects may rapidly increase to repopulate the originally infested areas and costs may again exceed benefits especially in fringe areas of the pest population when large acreages are infested at population densities too low to cause significant crop losses. Proponents of the goal to eradicate the boll weevil argue, however, that although a continuing surveillance and eradicative effort may be necessary in non-isolated populations like that of the boll weevil, overall costs are likely to be lower than the benefits.

Population models on which eradication programs are based often assume that the genetic constitution of the target population will remain constant during the eradication effort. But past experience has shown that the genetic plasticity of insects is such that a pest may become resistant to intensive use of chemical insecticides or other suppressive measures employed. Migration and dispersal of the target species and the presence of alternate hosts may well defeat eradication efforts. Specific problems associated with such programs and costs of possible failure are considered in the Report of the Executive Committee. Because of these known problems and perhaps other as yet unknown problems, responsible scientists disagree about the possibility of success of the major program. Proponents argue that the important advances have been made in detection and suppression methods, and that if the trial program proves successful the long-range economic benefits in relation to costs and the improvement in environmental quality would fully justify an eradication and containment effort. The Cotton Study Team, however, believes that although the individual technologies comprising the project are valuable tools for pest management programs aimed at containment, the effort to eradicate a major continental pest species is not likely to achieve the successes now predicted for it by proponents.

Systems analysis and computer simulation modelling of the major components of the cotton agroecosystem can be used to estimate the reaction of the system to various exogenous factors; to organize and coordinate data from experiments on different parts of the system; and to clarify what additional information is needed.

We recommend that further development of these techniques be encouraged and that they be used to coordinate the research efforts of entomologists, plant pathologists, and weed scientists and to coordinate the dissemination of information to farmers for control operations.

Within the cotton industry, there are a number of developments that, on the surface, appear to be peripheral to pest management problems but that could have a significant long-run impact on future pest control problems and practices. Developments within the production sector include high-density plantings, use of growth regulators, new seed varieties, and new harvesting and ginning techniques. New techniques in the processing and utilization of cotton in the textile industry are being developed. For example, the potential for cottonseed protein in human nutrition might allow the commercial production of radically different plant types with altered pest problems.

We recommend that future pest control research planning include consideration of these developments.

For the immediate future, the combination of various pest control practices--chemical, cultural, biological, and genetic--into programs of integrated pest management offers the most promising direction for U.S. cotton production. The successful development and delivery of integrated pest management programs will require the development and the introduction of new knowledge, as well as new material inputs into the production process. In some areas, the success of a pest management program may depend on the creation of new institutions for collective action which require that participation by cotton producers be compulsory. When management programs depend upon private individuals serving as consultants, legal questions concerning licensing and liability must be considered, as well as new economic incentives for the pest control industry.

Cotton is produced under such varied conditions by so many farmers that a typical cotton farm or farmer does not exist. Even within specific cotton production regions, only broad generalizations can be made about production patterns. It should be recognized that any given pest management program will achieve varying degrees of success among different economic classes of producers. Farm operators of different educational levels and of different tenure status will vary in their ability to implement sophisticated pest management programs. The profitability of certain pest management practices may depend on the size of the farm. Although the widespread adoption of pest management programs throughout the Cotton Belt would lower aggregate cotton production costs, the magnitude of such cost reductions would not be uniform among all producers.

We believe it important that questions of economic and social equity be considered when new programs of pest management are introduced.

In addition to the general recommendation contained in this summary, the following are specific points that the study team feels need immediate attention.

1. Government farm programs based on price supports and acreage controls have in the past provided incentives for heavy use of pesticides on cotton. While not taking issue with the objective of these programs, we recommend *that if similar policies are considered in the future, consideration be given methods that will achieve the same goals without continued heavy pesticide use.*

2. The role of land tenure relationships in determining pest control procedure in the Cotton Belt is an important area in which research is needed. We recommend *that funds be allocated for research in this area.*

3. Although examples of successful adoption of improved pest control technology have been illustrated, *stronger emphasis needs to be placed on research into the systems for delivering new pest control technologies to farmers.*

4. *A larger fraction of federal research funds allocated to pest control should be devoted to pest management.*

5. It is also important to *focus more research activity on the legal structures and incentives needed to ensure the continued development of the pest management industry.*

6. *An additional effort is also needed to train individuals engaging in pest control activities, whether they are farmers or consultants.*

7. *The present level of research on the environmental impact of currently used pesticides and their potential substitutes is inadequate.*

1

INTRODUCTION

Cotton is a major crop in U.S. and world agriculture. It plays a dominant role in the economy of the U.S. Cotton Belt and traditionally accounts for over 50 percent of the income from the sale of crops in at least 10 states. The Cotton Belt has as northern boundaries central California, central Arizona, New Mexico, Oklahoma, southern Missouri, southern Illinois, Kentucky, and North Carolina, and the southern boundaries are Florida, the Gulf of Mexico, and Mexico.

Cotton is produced under a variety of environmental conditions. However, requisites for its propagation are a long growing season, fertile soil, warm temperatures, and adequate water. All cottons belong to the genus *Gossypium*. Only four species are cultivated in the world, and only two in the United States. Upland varieties, *G. hirsutum* L., are grown on about 99 percent of the U.S. acreage and almost 90 percent of the world acreage. These varieties are grown as annual subshrubs characterized by short to medium, relatively coarse fibers produced in four or five loculed capsules (bolls). The American-Egyptian (pima) types, *G. barbadense* L., are grown on the remaining U.S. acreage. These are characterized by long, fine fibers and small to medium bolls with three or four locules.

ASPECTS OF THE COTTON ECONOMY

A major consideration in the profitable production of cotton is an efficient and timely pest control program. Because of the wide range of soil and climatic conditions under which cotton is grown, no one set of pest control practices can be recommended for all the cotton growing areas of the United States. The situation is further complicated by the year to year fluctuations in the profitability of the cotton industry. The price of cotton has risen and fallen in response to bad weather; insect infestations; war; the expansion of acreage into new areas; new production technologies; government support programs; and, more recently, competition from synthetic fibers. When prices are high, acreage expands into marginal production areas and farming practices, including pest control, are intensified. When prices are low, acreage is reduced and farming practices relaxed. The intensity and rapidity of the changes in market signals is illustrated by the increase from 31 cents to 90 cents per lb of lint for U.S. cotton between September 1972 and September 1973 in the world (Liverpool) market. The United States has established adjustment and incentive programs to shelter domestic cotton producers partially from these disruptions. But agricultural institutions also change, and such changes have affected the location and intensity of production practices as directly as have changes in prices.

Cotton's future in the United States is also uncertain because the direction of cotton price changes on the world market is unclear and future U.S. cotton policies are unknown. This uncertainty is intensified by a foreseeable shift in the role of cotton. Historically, the presence of gossypol (a toxic dimeric sesquiterpene) in cottonseed has prevented human consumption of cottonseed protein. Processes now exist to remove the gossypol economically. Shifts to greater seed-yielding varieties of cotton might result in a higher level of protein production per acre than is currently obtained from soybeans. In addition, once gossypol is removed, cottonseed protein is more bland than soybean protein and, therefore, more adapted to processing with other foods and flavorings. Thus, there is potential for a dramatic change in the reasons for growing cotton and, hence, in the production practices utilized.

This report could become obsolete in a very few years if recommendations were directed only to a future based on projections of the recent past. It is important to recognize and make recommendations in light of the probable range of future economic environments in which cotton might be produced. For the purposes of this study, three principal economic environments are considered. The first projects moderate prices for U.S. farmers through continuation of world price trends and maintenance of income support and acreage control programs generally comparable to those of the past. The second projects low prices resulting from the removal of government programs and the continuation of the historic decline of cotton prices in the world market. Under these conditions, significant quantities of cotton would be grown only in low-cost production areas, such as the Mississippi Delta and parts of California and Texas. This constriction of cotton acreage would reduce the problem of pest management on cotton generally and would reorder research and management strategies toward the major pests in low-cost areas. Also, a decline in cotton prices might reduce the economically optimal level of pest management because, under this scenario, the costs of control would be relatively higher than the decreased value of cotton lost to pest damage. The third economic projection is based on potential impact of high cotton prices. With high prices, such as those in 1973-1974, cotton acreage may expand into areas that are now submarginal.* The total set of pest management problems would increase for two reasons: first, problems would increase due to the expansion in acreage and, second, with high prices, the relative cost of pest control to the value of crops lost would decline. Thus, more intensive pest management would become economically optimal. Pest management strategies within this scenario would undoubtedly differ depending on whether the high returns resulted from high lint prices or from high seed prices because of the adoption of the gossypol removal process.

The effects of projected economic conditions on cotton production are complex. Each of these conditions can influence regional production patterns and pest population characteristics, thereby influencing the environmental impacts and institutional requirements of pest management, the applicability of certain pest management practices, and profitability of the other production and harvesting techniques.

---

*Depending on the relative prices of competing crops, cotton could expand into additional high production acreage as well. For example, in areas of the Mississippi Delta where cotton competes with soybeans, higher cotton prices could result in expansion of cotton into soybean acreage.

2

FACTORS AFFECTING COTTON PRODUCTION PRACTICES

MARKET CONDITIONS AND GOVERNMENT PROGRAMS

In 1972, 19 states, from North Carolina to California, produced 13.6 million bales of cotton on 12.8 million acres. A map of cotton acreage in 1969 is shown in Figure 2-1. One hundred years ago cotton was produced only in the southern and border states. At that time, the crop was 3.9 million bales harvested from 9.6 million acres. During the intervening years, major changes occurred in the cotton sector of U.S. agriculture. Cotton growing areas spread westward and total acreage expanded until 1926. Since then a downward trend in total harvested acreage has appeared as the relative importance of cotton production in the southeastern states slowly declined. Although wide variations in the annual production of cotton have occurred, the general production trend has been upward. With declining acreage devoted to cotton, the level of production has been maintained through increased yields. These trends in acreage, production, and yields are shown in Figures 2-2, 2-3, and 2-4.

From the end of the Civil War to the late 1930s, per acre yields remained quite stable, averaging slightly more than 150 lb of lint per acre. Since then yields have increased steadily, leveling off at approximately 500 lb per acre in the mid-1960s. A number of factors were responsible for increased yields. Acreage utilization has shifted to the more efficient (lower cost) cotton producing areas and producers. Although the major regional shifts in cotton producing areas occurred prior to 1960, within regions the more efficient farmers continually produce a

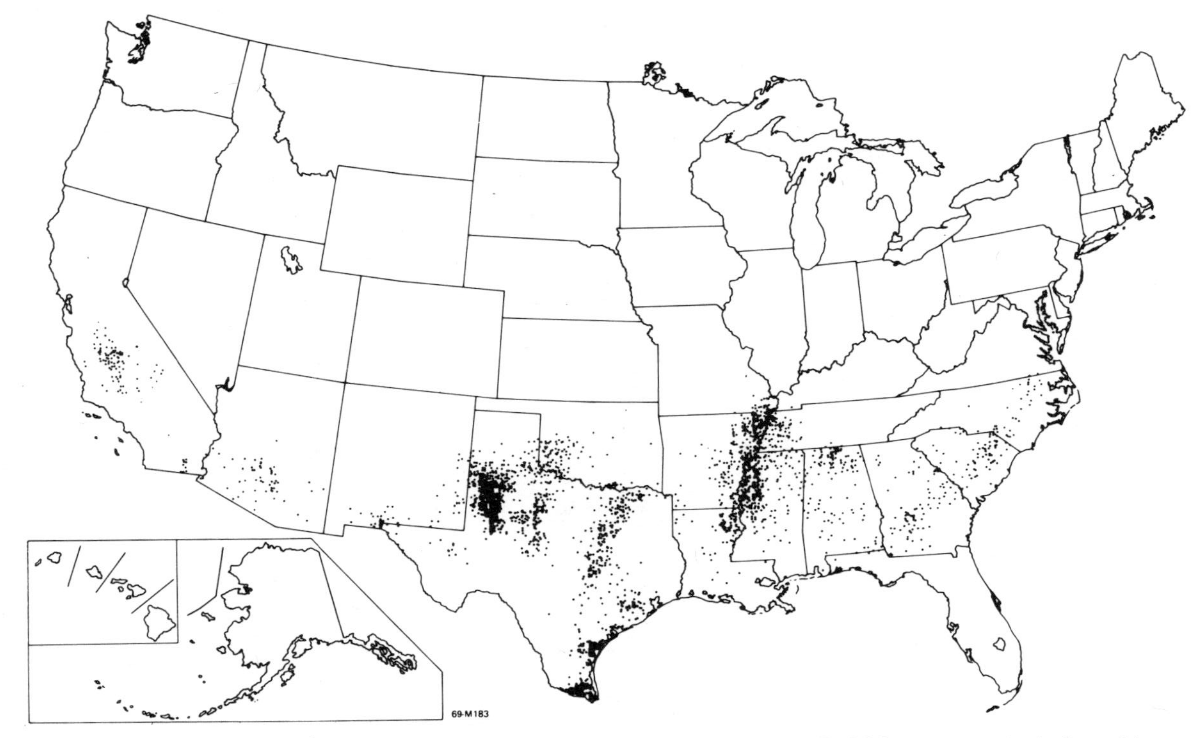

FIGURE 2-1 Cotton acreage, 1969. Each dot represents 5,000 acres; total cotton acreage in the United States was 11,496,320 acres. (Source: Department of Commerce, Social and Economic Statistics Administration, Bureau of the Census.)

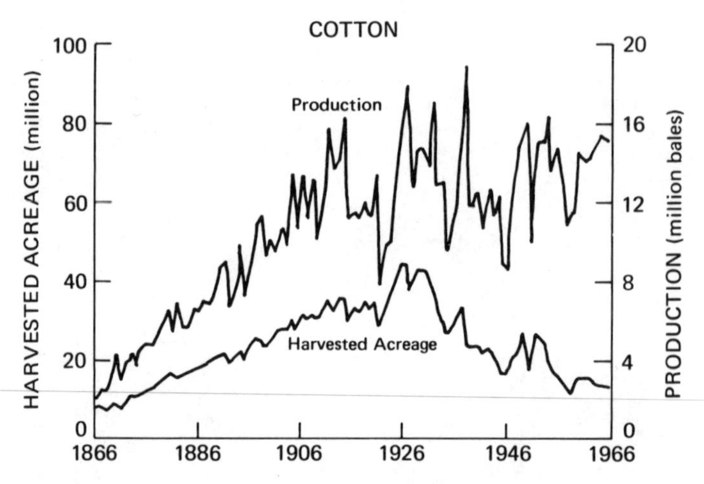

FIGURE 2-2 Cotton acreage and production. (Source: U.S. Department of Agriculture, Statistical Reporting Service.)

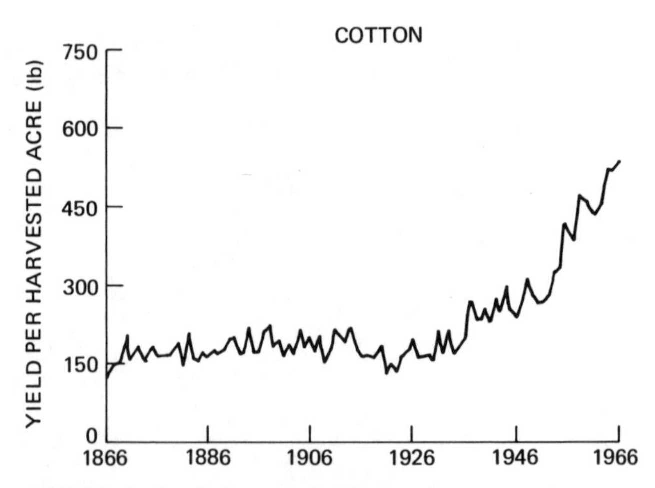

FIGURE 2-3 Cotton yield per harvested acre. (Source: U.S. Department of Agriculture, Statistical Reporting Service.)

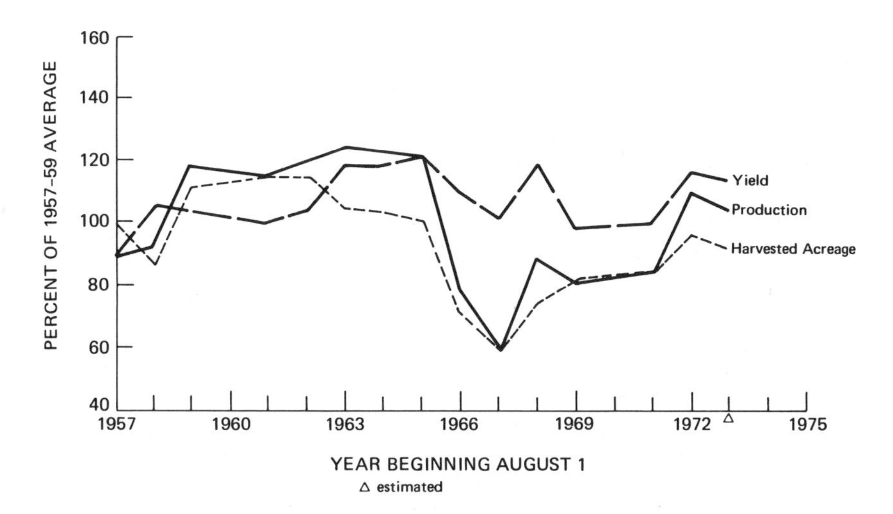

FIGURE 2-4 U.S. cotton acreage, yield, and production. (Source: U.S. Department of Agriculture, Economic Research Service.)

larger portion of the crop, thus increasing overall yields.* Irrigation and fertilizer use also have made significant contributions to increased yields. By 1964, three-fourths of the cotton acreage received fertilizer applications. The continuing development of improved cotton varieties has resulted in cotton plants with yield-enhancing characteristics, such as disease resistance.

Over the past three decades, the development of synthetic organic insecticides has allowed cotton producers to reduce losses to primary cotton insects. Initially, the use of these insecticides was an important factor contributing to increased yields. However, the yield-increasing impact of insecticide use has been modified in recent years. Resurgent populations of primary pests, increased problems with secondary pests, and development of resistance

---

*A cultural practice significantly affecting yields has been planting cotton in skip-row patterns. Skip-row planting is the practice of alternating two or more rows of cotton with idle land. This practice increases yields by allowing the plant access to additional moisture and additional room to grow and mature. It also increases reported yields because only those rows planted to cotton are counted as cotton acreage.

to insecticides are problems that have mitigated the beneficial results of increased insecticide use.

The use of synthetic organic herbicides in cotton production has reduced competition from weeds and has been a significant factor reducing the land preparation, labor, cultivation, and amounts of water needed. Historically, the major labor demands in cotton production have been for weed control and hand picking. The reduction in labor required for weed control has complemented the reduced labor needed in cotton picking. With the use of mechanical harvesters, hand picking harvests less than 1 percent of the cotton crop. Similarly, the general acceptance of chemical weed control has reduced the major labor requirement during the growing season. At present, less than 30 labor hours are required to produce one bale of cotton. Twenty years ago, about 120 labor hours were required.

Today, cotton is produced on approximately 300,000 farms. Perhaps one-third of these farms plant less than 10 acres of cotton each year. Production practices, costs, and yields vary considerably across the Cotton Belt. However, in 1969, the USDA attempted to measure the inputs used in producing cotton and to determine their effect on yields (Starbird and French 1972).* The Cotton Belt was divided into three regions and twenty specific areas having similar production characteristics (Figure 2-5).

Cost concepts used in the survey were variable costs, direct costs, and total costs. Variable costs include only those items used directly in crop production. These expenses which vary with production year include hired labor (except hired overhead labor), operating costs of machinery and equipment, costs of all materials used, and interest on operating capital. Direct costs include variable costs plus unpaid labor, hired overhead labor, and depreciation and interest on machinery and equipment. These expenses vary less within a crop year. Total costs include direct costs plus all allocation of general farm overhead costs and land charges. They exclude a return for unpaid management.

For analytic purposes, total costs indicate the level of return necessary for continued production over a long period of time. Direct costs permit comparison of cotton's competitive position over an intermediate time period.

---

*Abnormal weather conditions affect yields and costs per unit of output. Unfortunately, yields among the sample farms were relatively low in 1969 due to unfavorable weather conditions in many production areas.

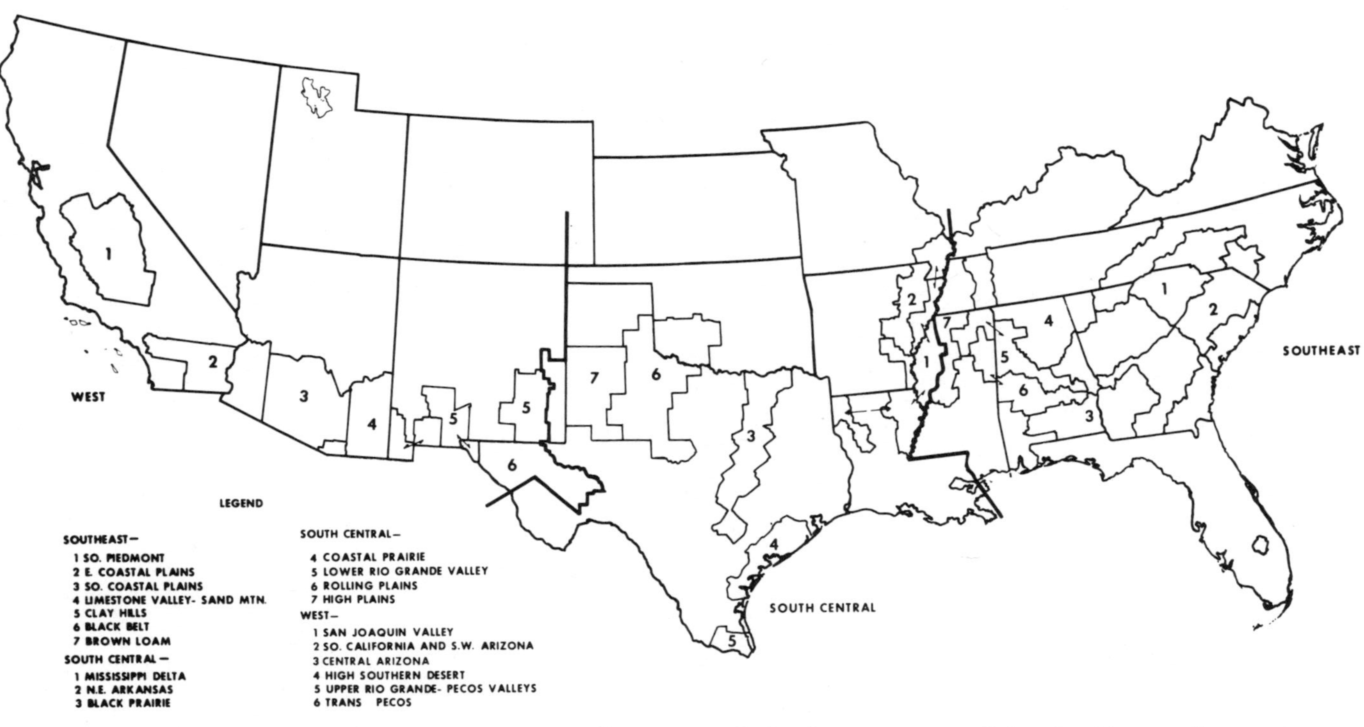

FIGURE 2-5 Cotton production regions. (Source: U.S. Department of Agriculture, Economic Research Service.)

Variable costs provide data for annual decisions on what and how much to produce.

Costs varied widely between regions surveyed. Total costs per pound of lint were as high as 46.5 cents in the Southern Coastal Plains and as low as 26.3 cents in the Rolling Plains (see Table 2-1). Direct costs per pound showed similar variation. Yields and costs per harvested acre were highest in Southern California-Southwest Arizona (1,110 lb and $415) and lowest in the Black Prairie of Texas (176 lb and $77). In three areas of the Southeast (Southern Piedmont, Eastern Coastal Plains, and Southern Coastal Plains) and in Central Arizona, total costs per pound exceeded total receipts per pound (market price plus governmental direct support payments). Variable costs per pound of lint in each area were lower than total receipts per pound. However, in the three high-cost areas of the Southeast, average variable costs were higher than market prices received for lint. Table 2-2 shows the average cost per acre for the major inputs in U.S. cotton production. In 1969, the average cost for pesticides (herbicides, insecticides, defoliants, and other chemicals) was 9.7 percent of direct costs and 7.6 percent of total costs per harvested acre.

Government Programs

Since the 1930s, public programs designed to support and protect domestic farm incomes have had direct impact on cotton production in the United States. In addition, these programs have influenced world production of cotton and competing fibers. Although the specific provisions of U.S. domestic farm programs for cotton have been modified over the years, generally they have included some type of price support and acreage control mechanisms.

Prior to 1965, domestic cotton prices paid to the producer were maintained by price support loans. Participating farmers were given the option of selling their crop at prevailing market price or of obtaining a nonrecourse loan from the USDA's Commodity Credit Corporation using their crop as collateral. The Corporation stood ready to loan the producer a specified amount per pound of cotton (known as the loan rate). If prevailing market prices were unfavorable at harvest time, a producer could place an entire crop under the loan program. Subsequently, the producer could opt to repay the loan principal plus interest, or, at maturity, to surrender the commodity securing the loan in full satisfaction of principal and interest. Thus,

TABLE 2-1 Average Yield of Upland Cotton and Production Costs per Acre and Production Costs and Receipts per Pound of Lint, U.S., 1969

Region[a]	Yield per Acre Harvested (lb)	Cost per Acre Harvested ($)		Costs and Receipts per Pound of Lint (¢)			
		Direct	Total	Variable Costs per Pound	Direct Costs per Pound	Total Costs per Pound[b]	Receipts per Pound[c]
Southeast							
Southern Piedmont	421	145.40	179.61	22.5	30.4	37.6	37.4
Eastern Coastal Plains	442	182.30	214.17	27.0	36.7	43.1	39.5
Southern Coastal Plains	393	175.95	202.60	28.9	40.4	46.5	40.2
Limestone Valley Sand Mountain	457	128.05	165.49	17.6	24.5	31.6	35.9
Clay Hills	495	145.50	184.23	18.8	25.8	32.7	36.7
Black Belt	437	146.75	176.92	22.7	29.6	35.7	36.6
Brown Loam	519	140.00	178.57	16.8	23.4	29.8	34.7
South Central							
Mississippi Delta	559	144.84	185.98	16.7	22.4	28.7	35.0
Northeast Arkansas	496	128.93	165.33	16.4	22.4	28.7	32.6
Black Prairie	176	59.24	76.84	20.1	30.1	39.0	41.7
Coastal Prairie	294	83.93	109.57	17.4	25.3	33.1	36.5
Lower Rio Grande Valley	512	144.21	183.90	19.8	24.9	31.7	33.7
Rolling Plains	284	65.04	89.21	13.4	19.2	26.3	34.5
High Plains	339	96.99	126.00	17.0	24.4	31.7	35.7
West							
San Joaquin Valley	818	236.37	313.65	19.5	25.0	33.1	38.7
Southern California-Southwest Arizona	1110	338.06	414.94	22.5	26.6	32.7	34.0
Central Arizona	955	318.24	389.32	23.7	29.5	36.1	34.5
High Southern Desert	843	266.32	330.64	18.2	27.4	34.0	35.0
Upper Rio Grande-Pecos Valley	603	186.91	252.02	18.6	27.0	36.3	38.1
Trans Pecos	642	269.27	333.92	28.8	37.3	46.3	43.5
United States, average	455	131.18	167.93	18.5	25.0	32.0	36.0

[a] See Figure 2-5, page 9, for location of regions.
[b] Value of seed subtracted from total costs of producing lint and associated seed.
[c] Includes support payments.

SOURCE: Starbird and French 1972.

TABLE 2-2 Average Costs per Acre in U.S. Cotton Production

Region[a]	Yield per Harvested Acre (lb)	Average Cost per Harvested Acre ($)										Total Cost per lb of Lint (¢)[c]
		Labor	Power and Equipment	Materials: Fertilizer	Materials: Insecticides	Materials: Total	Other Direct[b]	Total Direct	General Overhead	Land	Total	
Southern Piedmont	421	21.63	49.89	23.20	11.51	47.45	26.43	145.40	20.38	13.83	179.61	37.6
Eastern Coastal Plains	442	22.42	64.11	28.61	18.22	61.37	34.40	182.30	13.37	18.50	214.17	43.1
Southern Coastal Plains	393	16.32	66.63	24.22	16.19	55.98	37.02	175.95	9.43	17.23	202.60	46.5
Limestone Valley-Sand Mountain	457	14.79	48.17	18.21	8.06	37.20	27.89	128.05	13.32	24.11	165.49	31.6
Clay Hills	495	19.75	49.74	16.56	6.74	33.61	42.40	145.50	21.33	17.40	184.23	32.7
Black Belt	437	22.90	49.75	19.33	12.40	43.31	30.80	146.75	14.63	15.54	176.92	35.7
Brown Loam	519	21.09	52.34	16.21	4.78	31.28	35.30	140.00	14.64	23.93	178.57	29.8
Mississippi Delta	559	22.03	50.85	11.54	10.92	35.61	36.34	144.84	15.27	25.86	185.98	28.7
Northeast Arkansas	496	22.25	48.01	11.57	2.41	24.62	34.05	128.93	11.14	25.26	165.33	28.7
Black Prairie	176	11.59	22.71	7.32	2.40	14.65	10.29	59.24	6.37	11.23	76.84	39.0
Coastal Prairie	294	12.78	33.62	8.78	3.27	19.60	17.93	83.93	8.49	17.15	109.57	33.1
Lower Rio Grande Valley	512	25.95	39.37	8.33	11.10	28.20	50.69	144.21	12.61	27.08	183.90	31.7
Rolling Plains	284	12.59	23.24	1.51	1.26	8.32	20.88	65.04	7.33	16.84	89.21	26.3
High Plains	339	17.98	29.38	5.91	0.53	15.52	34.10	96.99	9.91	19.10	126.00	31.7
San Joaquin Valley	818	50.51	62.96	15.37	10.92	37.59	85.30	236.37	29.04	48.25	313.65	33.1
Southern California-Southwest Arizona	1,110	53.88	69.10	28.35	35.99	82.01	133.06	338.06	29.44	47.44	414.94	32.7
Central Arizona	955	51.38	64.68	17.47	21.81	51.41	150.78	318.24	29.73	41.35	389.32	36.1
High Southern Desert	843	54.96	89.32	13.47	3.01	25.06	96.98	266.32	26.37	37.94	330.64	34.0
Upper Rio Grande-Pecos Valleys	603	40.69	66.07	8.26	3.38	20.20	59.95	186.91	26.01	39.10	252.02	36.3
Trans Pecos	642	48.03	51.63	18.74	14.39	42.82	126.79	269.27	35.14	29.51	333.92	46.3
United States, average	455	21.97	42.46	10.90	6.79	27.83	38.93	131.18	13.64	23.11	167.93	32.0

[a] See Figure 2-5 for regions.
[b] Includes the cost of irrigation, ginning, custom services, and interest on operating capital.
[c] Obtained by subtracting the value of seed from the total costs of producing lint and associated seed and dividing by yield plus an allowance for the weight of bagging and ties.

NOTE: Totals do not necessarily add because of rounding.

SOURCE: Starbird and French 1972.

because Commodity Credit Corporation loans were readily provided to producers, the loan rate effectively set a minimum level for domestic cotton prices. Through this procedure, high domestic prices were maintained and government inventories of surplus cotton were accumulated. Because the United States dominated world cotton production during this period, the U.S. domestic cotton policies effectively maintained world cotton prices at higher levels than would have prevailed otherwise. This encouraged foreign production and competition from synthetic fibers.

The cotton program written into the Agricultural Act of 1965 provided lowered price supports, direct payments to producers, and no subsidies to domestic users or exporters (Firch 1973). Prior to 1965, cotton prices were held at high levels by high loan rates. In recent years loan rates (and, therefore, market prices) have been allowed to move to lower levels while producers' incomes were maintained by the direct payments. The amount of the direct payment, prior to payment limitations, was the sequential product of allotted acres multiplied by a specifically defined base yield multiplied by the price differential as described below. This procedure effectively maximized the only element of the equation controlled by the producer.

The 1973 Agriculture and Consumer Production Act continued the trend toward allowing domestic cotton to be traded near world price levels. The Act established a target price of 38 cents per pound for the 1974-1975 crop years and linked loan rates to world cotton price levels. Thus, if prices fall below 38 cents per pound, a direct price support payment will be made to producers. The amount of the payment per pound of lint will equal the difference between the target price and the higher of either the loan level or the average market price received by farmers. This payment per pound of lint multiplied by the grower's base yield per acre multiplied by the acres in the grower's allotment all determine the size of the payment due the grower. Prior to 1970, direct payments of over a million dollars were made to a few individual growers. In 1970, direct support payments were limited to $55,000 per producer. The 1973 Agricultural and Consumer Act lowered the limit to $20,000 per producer. Since 1965, provisions also have been added to increase support payments to farmers with 10 acres or less of cotton.

In order to be eligible for these loans and payments, cotton producers have operated under complicated acreage control systems based on acreage allotments. The objective has been to achieve a national production roughly equal to domestic, export, and carryover needs.* A national cotton acreage allotment was apportioned to states, counties, and farms on the basis of historical production records. Typically, eligibility for loans and payments required that certain acres be removed from cotton production. To receive price support loans and payments, the farmer was required to plant within the acreage allotment. In fact, between 1954 and 1970 stiff marketing penalties prevented domestic cotton production outside of the program. Since 1970, cotton producers have been required only to set aside a certain percentage of their allotment. The producer then becomes eligible for loans based on total production and for payments on acreage planted within the allotment, even if more than the allotted acreage was planted. The 1973 Act retains the set-aside feature, but for the 1974 crop year there is no requirement for set-aside acreage.

One effect of acreage controls has been to restrict interregional and interfarm adjustments in the land resources devoted to cotton. Allotments are not transferable across state lines. However, allotments may be transferred between farms and counties within a state. Since 1971, farmers have been allowed to plant cotton without an allotment, but such production has not been eligible for price support payments. At present, nonallotment acreage can be brought into cotton production, but unless an allotment is purchased or leased from another producer in the same state or county, the new acreage does not develop an allotment history. Conversely, failure to plant at least 90 percent of an acreage allotment will reduce the following year's allotment, and the entire allotment can be lost if no cotton is planted for three consecutive years.

Delivery and implementation of advanced technology has been impeded by government programs which stress maximum yields per acre. Advanced technology, especially in the area of pest management, may slightly lower unit yields while increasing net profit. Such technology has been ignored by some cotton producers because a lower base yield would earn a lower total direct support payment.

---

*Although no longer of major importance, the Soil Bank Act of 1956 and the Cropland Adjustment Program provision of the 1965 Food and Agriculture Act represented additional attempts to reduce production by taking farmland out of production. The Soil Bank Act has expired and only 138,000 acres (mostly in Texas and the Southeast) remained in the Cropland Adjustment Program in 1973.

Perhaps of greater significance to this report is the role maximum yield incentives play in the overkill approach to insect pest control. Some of the current environmental problems occupying this study team have been created by growers willing to pay for additional inputs in order to increase yields. These short-term advances have, in some cases, evolved into unit yield reductions because of the destruction of beneficial insects and/or the ensuing resistance of insect pests to insecticides used.

An alternative government program which allots *production* rather than *acres* should deemphasize yield per acre and encourage delivery and implementation of advanced technology in all production areas.

Export Markets

Cotton has always been a major export crop. Although the volume of exports varies, historically, cotton has been the fifth most valuable export crop in the United States.

World production of cotton was approximately 55 million bales in 1972. Major producing countries are the United States, the Soviet Union, China, India, Pakistan, Brazil, Egypt, Turkey, Mexico, and Sudan. By exporting about 36 percent of its crop each year since 1965, the United States has remained a major exporting country. Traditionally, the United States has been a residual supplier of cotton in foreign trade; that is, the difference between foreign production and foreign consumption has been met from U.S. production.

During the period of high domestic prices and surplus production in the mid-1950s, two government programs were initiated to expand exports. The Agricultural Trade Development and Assistance Act of 1954, (PL 480), was designed to stimulate exports of surplus agricultural commodities by sale of cotton (and other commodities) for foreign currencies or long-term dollar credits. Exports from both government and private inventories have occurred under this program. Between 1955 and 1966, the United States provided an export subsidy. The subsidy paid to exporters roughly equaled the difference between the U.S. domestic market price and world prices, allowing U.S. cotton to enter the export trade at world prices.

EXPERIENCE AND EDUCATION OF PRODUCERS

The experience and educational levels of cotton producers exert an important influence on production practices. The increasingly sophisticated production technologies being introduced to modern U.S. agriculture often require that the user have some rudimentary understanding of underlying scientific principles. This is especially true of pest management technology being developed for cotton. Thus, the experience and educational levels of cotton producers may significantly influence the adoption of new pest control practices and different sets of pest management practices may have to be developed for producer groups with differing levels of education and experience.

Although differences are decreasing, farm operators and workers have had a low level of education in comparison to the non-farm labor force. These lower educational levels may be related partially to the age distribution of the farming population. The 1969 Census of Agriculture indicated that over 65 percent of the cotton acreage was harvested by operators over 45 years of age. Farmers under 35 years of age harvested only 12 percent of the cotton acreage. Furthermore, the distribution of education within the farm population may be bimodal. Farm operators near the poverty level and hired farm workers would be clustered at low educational levels while other farm operators would cluster at rather high levels of education. The widespread adoption of improved agricultural production techniques (mechanization and chemical pest control, among others) has increased the relative value of the highly skilled and educated portion of the farm labor force. It also appears that over the long run, increases in the level of schooling and increases in the combined research and extension effort have increased the inequality of farm incomes (Gardner 1969).

INFORMATION SOURCES

Cotton producers rely on numerous sources for information on new farming practices and techniques. While obtaining information on new farming practices from the media, producers also may rely on extension agents, pesticide dealers, pest control consultants, and other farmers for information on cotton pest control problems.

Extension Agents

The agricultural extension services, working through county agents, have played a major role in delivering knowledge of new agricultural technologies to farmers. The process has involved a flow of information between farmer, extension service, and research institution. Perceptions of problems are transmitted from the farm level through the extension service to researchers. Similarly, the extension services translate the products of agricultural research into forms useable by the farmer. At the farm-extension interaction level, traditional extension methods include farmer program development committees, farmer meetings, demonstrations of new practices, newsletters, and news releases. In addition to programs aimed directly at improving agricultural management practices, the extension services conduct many programs in the areas of natural resource management, home economics, community improvement, and youth development (the "4-H" program).

The extension approach has generated many successful pest control programs. The process usually involves educating the farmer: (a) to be aware that a potential pest problem exists; (b) to be able to identify specific pests on the farm; and (c) to know the specific control actions to be taken. The most successful of these programs, whether the target was an insect, weed, or disease pest, have required little evaluation of damaging pest levels.* That is, the remedial action (application of insecticide, application of herbicide, use of treated seed) did not require the farmer (or county agent) to make complicated management decisions based on interacting factors such as climatic conditions, biology of the pest species, or presence of populations of beneficial insects.

Many state extension services have been unable to provide the intensive technical assistance needed by individual cotton growers to make the frequent insect population assessments required for precise timing of insecticidal treatment. The typical county agent simply lacks the time and entomological training. As a result, many successful extension programs have educated farmers to perform

---

*Illustrative examples are cover spray programs for fruit and nut crops. Insecticidal sprays are made prior to bloom, during bloom, post bloom, etc. The decision to begin treatments is made on the basis of the stage of fruiting of the tree and/or the first appearance of the insect pest.

relatively straightforward pest control activities. Extension programs requiring complicated pest management decisions by farmers (or requiring readily available sophisticated technical assistance) have been less successful.

However, the extension services have begun to develop new types of pest management assistance programs. The goal of these pilot programs has been to establish pest control operations that use both chemical and nonchemical methods in an effective, efficient, and safe manner (USDA 1973a). The pilot programs for cotton are designed to provide participating farmers with the timely information and advice necessary to make better pest management decisions. Paraprofessional "scouts" conduct frequent surveillance operations in the participants' fields. Extension professionals and research specialists at cooperating institutions use these scouting reports to formulate advice to the farmer on pest control operations. By 1973, 14 states had initiated this type of cotton pest management program. Although it is too early to evaluate the long-term potential of this approach, one analysis showed that in Texas the net return to the producers over a three-year period was $6.39 for each $1 invested (Frisbie et al. 1974).

Pesticide Dealers

Businesses retailing pesticides to farmers are a diverse group. Some dealers may sell only pesticides, whereas others may function primarily as a general farm supply agent, a commercial applicator, an implement dealer, or a cotton ginner. Therefore, incentives to assist farmers may vary.

In addition to supplying farmers with the chemical inputs for pest control, the local pesticide salesman or dealer serves as a source of information on pest control. Most farmers expect the pesticide dealer to provide information on the efficacy and proper use of pesticides. In addition, farmers often expect the dealer or salesman to identify their pest problems and to recommend specific remedial chemicals (Beal et al. 1969).

Pest Management Consultants

Pest management consultants serve as an additional source of pest control information. Consultants are private entrepreneurs who provide pest management expertise on a fee-per-acre basis. Generally, consultants make a greater effort through pest, predator, and plant vitality assessment to determine the real need for pesticide applications. When the pest consultant industry developed in the early 1950s, cotton was among the first of the crops served. In California, 19 consulting firms, working at least part-time on cotton, now provide pest management advice on about 151,000 of the 868,000 acres planted to cotton in the state (Norgaard and Levinson 1974). Pest management consulting services are also being established in other Cotton Belt states.

The typical consultant is well-trained in the biological sciences or in entomology. A consultant often employs less well-trained individuals as scouts to check the fields. When pest populations increase, the surveillance frequency increases and the consultant begins to assess and discuss these developments with the farmer. The consultant estimates whether the infestation will be kept below critical levels by parasites, predators, or other forms of natural control and recommends specific control measures. Initiating or delaying irrigation may be recommended to affect the temperature, plant vitality, or rates of development of both predator and prey. Treatment of only a portion of the field may be recommended. If the situation appears especially unstable, the consultant may return to reassess the situation soon. A few consultants occasionally recommend the introduction of biological controls. However, the consultant's service can be described as insurance through surveillance and prediction rather than through the application of pesticides or their biological substitutes. Eliminating or delaying the application of pesticides increases the opportunity for natural controls to work in the future. Consequently, common problems related to secondary pest outbreaks and the further need for chemicals can be greatly reduced by consultants. Pest management consultants have effectively reduced the overuse of pesticides, but have only begun to recommend practices which resemble complex integrated control strategies.

The use of pesticides and total pest management costs are typically lower for consultant-supervised farms than for farms following salespersons' recommendations. In the San Joaquin Valley of California, total pest management costs for cotton under the supervision of consultants averaged $13.08 per acre including the consultant's fees, as compared to $15.57 per acre for cotton served by salespersons (Norgaard and Levinson 1974). Insecticide use during these two years was reduced by more than half when a consultant's advice was followed. Farmers who utilized consultants, however, used considerably more herbicides. Cotton acreage under consultant management averaged $35 greater yield per acre. The extent to which this greater yield can be attributed to the role of the consultant versus the overall skills of the farmer utilizing the consultant is unclear.

Pest management consultants face a number of operational problems. Except for consulting on citrus pest management, the work is highly seasonal. The length of the season can be extended by diversifying to cover several crops with different peak activity periods, but this solution entails broader expertise for both the consultant and scouts. Also, diversification frequently increases time spent traveling between growers and between different production regions. The consultant may keep reasonably busy in the slack seasons recruiting prospective clients, reassuring old clients, and studying recent research findings. However, there is not sufficient work to employ scouts, as well as laboratory and clerical assistants, in the slack season. "Vacationing" schoolteachers and students can be effectively used from mid-June to late August. However, most consultants need help both earlier and later than this. Because of the seasonal limits of the scout's role, consultants have not been able to develop a stable core of skilled workers who are recognized and trusted by the growers from year to year or even from the beginning to the end of the season. Trust in the monitoring and decision-making personnel is essential to the success of all pest management information delivery systems. This problem is most acute for the small, independent pest management consultant.

Many consultant firms originated as one-professional operations because of the unique characteristics of those who founded the industry and because a small number of growers in any region have been willing to experiment with new techniques. The industry is now in a period of transition when increasing demand for service has made growth and change possible. Expansion can mean adding the expertise of a professional plant pathologist, weed scientist, and perhaps an agronomist to firm decision-making capability. Expansion can support more sophisticated laboratory analyses and administration. Expansion, however, requires capital. To date, banks have been hesitant to lend money for expansion of business operations they do not fully understand. As a consequence, most growth has occurred at a gradual rate through internal funding.

A third problem, again related to professional trust and firm size, is that of insurance. A single consultant can provide pest management advice on several million dollars' worth of crops each year. To offset potentially crippling malpractice suits the consultant relies upon conservative approaches to pest management and more intensive pesticide application than if such threats were not present. Capable professionals who are unwilling to accept this risk will not enter pest management consulting. Insurance companies have not been interested in providing liability insurance for consultants, presumably because the field is new and the risks have not been established. At least one firm is now experimentally offering coverage limited to $10,000.

The pest management consulting industry also faces special problems related to basic differences between delivering knowledge and marketing material inputs. The consultant's product--knowledge of when to delay the use of pesticides--can be acquired and used independently by intelligent farmers, employees, and competing pesticide dealers. This filtering of knowledge speeds the implementation of better pest management practices, yet simultaneously erodes each consultant's market and long-term profits. Potential consultants, realizing that their market may erode through education, may hesitate to enter the consulting business. These factors may restrict the consultant industry to fewer than the optimum number of participants. If private consultants are to constitute a significant portion of a future pest management delivery mechanism, consideration might be given to public subsidy and insurance programs to maintain the supply and income of consultants at optimal levels.

SOCIAL AND INSTITUTIONAL FACTORS

Land Tenure and the Labor Force

Historically, land ownership patterns have been an important factor influencing production practices and the rate of adoption of new technologies. Tenure agreements in a given geographical region usually have clearly determined which party supplies certain inputs and how the crop is to be divided. The introduction of a new agricultural production practice usually required adjustments in these tenure relationships (Street 1957). Under current pest control practices on cotton, tenure relationships seem to be relatively stable. However, just as mechanization and herbicides required adjustments in the land tenure system of cotton production in the South, evolving pest control practices may require or stimulate further adjustments. These adjustments in tenure relationships probably would involve renegotiation of the productive inputs and crop division. For example, if a pest management program is to use scouts for pest population assessment of cotton fields, landlords and tenants must agree on some division of the expense. If mandatory plow down and stalk destruction dates were part of a pest management program, tenure agreements would have to state which party was responsible for compliance. (According to the 1969 Census of Agriculture, 28.5 percent of U.S. cotton farms were owner operated, 41.8 percent were operated by part owners, and 29.7 percent were tenant operated.)

Over the past 20 years, the rapid transformations in agriculture have been associated with large out-migrations of rural populations from the cotton producing regions of the United States. Most of this population was engaged in agriculture, either as farmers or hired workers. (Cotton production was the principal farm enterprise, but tobacco and peanut production were important.) The mechanization of cotton production and adoption of chemical herbicides have replaced hand labor in the weeding and harvesting operations. It is not clear to what extent the labor force was "pulled" out of cotton production by more attractive opportunities elsewhere and to what extent the labor force was "pushed" out of cotton production by the adoption of lower-cost technologies.

During the 1950s, net out-migration from the nonmetropolitan areas of the South reduced the population by 4.3 million people. This reduction included 2.7 million whites and 1.6 million blacks. Between 1960 and 1970, out-migration of southern nonmetropolitan whites decreased to 123,000, while out-migration of southern nonmetropolitan blacks decreased only marginally to 1.4 million. Although the total southern nonmetropolitan black population decreased only 4.5 percent between 1960 and 1970, the total nonwhite farm population (of which 90 percent are blacks) decreased 64 percent (Beale 1973).

Less information is available on Mexican-Americans involved in cotton production. However, significant out-migration of the nonmetropolitan Mexican-American population did occur during the two decades. Out-migration increased in the 1960s in Texas, particularly in the heavily agricultural (cotton producing) Lower Rio Grande Valley.

It is unlikely that any significant portion of this population will return to rural cotton producing areas. Thus, pest management practices dependent upon a large pool of relatively unskilled (and often mobile) labor would be extremely difficult to implement.

Local Business Communities

As a business enterprise, cotton farming requires the development of numerous relationships between the farmer and individuals or groups who purchase his products or supply his productive inputs. This agricultural infrastructure is capable of exerting a strong influence on farm production practices. Cotton production today requires substantial investments in land, machinery, and purchased inputs. These investments almost always are financed by either long-term or production credit. Local financial institutions, not unreasonably, encourage borrowers to follow commonly accepted "good" agricultural practices. Thus, a local production credit association may require that borrowers follow certain recommended fertilization or pest control programs.

In past years, a cotton gin often served a small, relatively well-defined geographical area and exerted considerable influence over the cotton producers in that community. Recently, increased competition, organization of cooperative gins, and the development of high capacity gin machinery have significantly altered this influence on a local community. However, gins do continue to influence production practices through the services offered to their farmer customers. Many gins provide production credit and some offer volume purchasing services for their customers. Often gins offer pest control services, including custom application of pesticides.

REGULATORY FACTORS

Cotton Variety Control

Except for producers in the San Joaquin Valley of California, cotton growers in the United States are able to choose the varieties to be planted. However, varietal control has been the objective of numerous federal and state programs. The concept of one-variety communities began in the irrigated cotton areas of the Southwest (Cook 1911). One-variety communities were proposed as a method of obtaining higher prices and seed quality in the area. Later, one-variety communities were established in the Southeast specifically to improve pest control (McLendon 1938, Ware 1938). Early attempts to control the boll weevil emphasized growing early-maturing varieties along the northern limits of the Cotton Belt. An uncontrolled increase of these seed lots produced mixtures causing general deterioration of the fiber quality. Thus, the need for early-maturing varieties and the market advantage of one-variety production encouraged the adoption of local variety control.

Only the varietal control program of the San Joaquin Valley continues in operation today. The reasons other programs have been terminated are not clear. One reason may be related to the incentive characteristics of the seed production industry. In 1973, approximately 85 percent of the U.S. plantings used seed originating from commerical seed firms (USDA 1973b). These seed companies encourage grower acceptance of new varieties in which they have a proprietary interest.

Thus, the San Joaquin Valley is the only remaining area in the Cotton Belt where the one-variety concept is still practiced. Although implementing the one-variety concept produces cotton of uniform quality throughout an area, it may limit rapid introduction of a new variety desirable under cultural and phenological aspects of integrated pest management.

Governmental Programs for Pest Control

Governmental programs directed toward the control of specific agricultural pests have been in existence for many years. Control programs usually establish quarantines and buffer zones or enforce planting and harvesting regulations. These actions may be supplemented by more direct pest suppression activities including: spraying or fumigating of fields and farm equipment, insect trapping, and sterile male release.

Although there are 19 federal domestic quarantines, most programs operate at the state level. The Animal and Plant Health Inspection Service (APHIS) of the USDA imposes federal quarantines in cooperation with the states. At present, the only federal quarantine of a cotton pest is the Pink Bollworm-quarantine Number 52. (APHIS cooperates with state agencies to control boll weevil in Texas. APHIS had primary responsibility for operation of the pilot boll weevil eradication experiment discussed elsewhere in this report.)

At the state level, control programs may encompass the entire state, or only infested portions. Target pests are not limited to insects, and may include weeds and plant pathogens. In California, cotton control districts may be established to prohibit cotton growing in areas where it is determined "that such prohibition is necessary for cotton pest control" (West California Agriculture Code 6051-6084). Districts must be formed with the consent of a percentage of the growers in the proposed restricted area. In Texas, the Governor can proclaim a noncotton zone and declare it a quarantine area. Besides attempting to quarantine cotton pests, pest control districts have been encouraging chemical pesticide application by levying per acre assessments to finance such applications (Ariz. Rev. Stat. 3-331. 01-10, 1973 Supp.). In Texas, pest control programs can be funded by assessments approved through a referendum of producers.

Environmental and Human Safety Regulations

Traditionally, government regulation of pesticides has been directed toward limited consumer protection objectives. The primary consumer to be protected was the agricultural user. The wide-scale application of pesticides was assumed to be beneficial, and the user was assumed to be able to react rationally to information about a pesticide's impact. Thus, pesticide regulation was confined to labeling requirements designed to insure that the user had accurate information about the efficiency of the pesticide on target species, the impact on nontarget species, and health hazards to humans. For example, the Federal Insecticide, Fungicide and Rodenticide Act of 1947 (FIFRA) required: (1) registration of chemical pesticides prior to their sale or movement in interstate or foreign commerce; (2) prominent display of poison warnings on labels of highly toxic pesticides; (3) coloring or discoloring of dangerous white powdered insecticides to prevent their being mistaken for foods; (4) inclusion of warning statements on the label to prevent injury to people, animals, and plants; and (5) inclusion of instructions for use to provide adequate protection for the public.

Registration was relatively nondiscretionary as the standards to be met were fairly clear. Not until 1964 was a manufacturer prohibited from registering a challenged pesticide under "protest." Such consumer protection legislation ignored environmental problems such as the poisoning of nontarget organisms and the risks posed to future generations by the application of persistent chemicals whose total impacts were not known and could not be predicted. The DDT controversy focused attention on the environmental protection defects in pesticide regulation and this controversy formed the basis for the Federal Environmental Pest Control Act of 1972 (FEPCA), which supplements FIFRA (Rodgers 1970).

FEPCA extended federal jurisdiction to intrastate commerce and expanded control by labeling to require discretionary review of a proposed registration if the administrator "determines that, when considered with any restrictions imposed under subsection (d):

> (A) its composition is such as to warrant the proposed claims for it;
>
> (B) its labeling and other material required to be submitted comply with the requirements of this Act;
>
> (C) it will perform its intended function without unreasonable adverse effects on the environment; and
>
> (D) when used in accordance with widespread and commonly recognized practice it will not generally cause unreasonable adverse effects on the environment.
>
> The Administrator shall not make any lack of essentiality a criterion for denying registration of any pesticide. Where two pesticides meet the requirements of this paragraph, one should not be registered in preference to the other. . . ."

Another innovative section of the legislation allows the Administrator to classify pesticides for general or restricted use. A pesticide is to be classified for general use when the Administrator determines that the pesticide "when applied in accordance with its directions for use . . . or in accordance with widespread and commonly recognized practice, will not generally cause unreasonable adverse effects on the environment. . . ." If a determination is made that use as described above "may generally cause, without additional regulatory restrictions, unreasonable adverse effects on the environment, including injury to the applicator . . ." the pesticide must be classified for restricted use. Such pesticides must be used only for specified uses or under direct supervision of a certified applicator. Additional conditions for use may be imposed. Experimental use permits may be issued if the Administrator determines that use is necessary to accumulate information necessary to register a pesticide.

Jurisdiction over pesticide registration was transferred from the Department of Agriculture to the Environmental Protection Agency. The EPA does not now evaluate the efficiency of pest control by application of pesticides or compare pesticide application to other means of pest control. FEPCA carries forward the principle that a pesticide is entitled to registration if pre-established disclosure standards are met (Large 1973). Also, the legislation provides for certification of pesticide applicators by the states under a program approved by the Administrator. More detailed discussion of the legislation is included in the Executive Committee Report.

Other important factors influencing pest control practices are the worker health and safety regulations under the Occupational Health and Safety Act (OHSA) and federal and state pesticide control acts. The purpose of OHSA is to promulgate administrative health and safety standards to protect workers against material impairment of health or functional capacity from regular exposure to a hazard. Protection of farm workers and applicators from the harmful effects of contact with pesticides is a relevant consideration in pesticide registration and use control. Thus, the EPA and state pesticide control agencies have somewhat similar missions. The Federal Environmental Pesticide Control Act of 1972 makes the user liable for failing to follow the label, which for the first time includes farm workers safety standards and provides the basis for safety standard enforcement. The EPA and OHSA administration have cooperated on the establishment of standards so that the proposed standards are not inconsistent. Both OHSA and EPA specify protective clothing to be worn by farm workers and both control application reentry and preharvest and harvest entry times for specified chemical pesticides.

REFERENCES

Beal, G. M., J. M. Bohlen, and W. A. Fleischman (1969) Behavior studies related to pesticides; agricultural chemicals and Iowa agricultural-chemical dealers. Iowa Agric. Home Econ. Exp. Stn. Res. Bull. 139 (80 pages).

Beale, C. L. (1973) Migration patterns of minorities in the United States. Am. J. Agric. Econ. 55:938-946.

Cook, O. F. (1911) Cotton improvement on a community basis. U.S. Dep. Agric. Yearb., 1911. U.S. Department of Agriculture.

Firch, R. S. (1973) Adjustments in a slowly declining U.S. cotton industry. Am. J. Agric. Econ. 55:892-902.

Frisbie, R. E., R. Parker, W. Buxkemper, W. Bagley, and J. Norman (1974) Texas cotton management: annual report, 1974. Unpublished document. Texas A & M University.

Gardner, B. L. (1969) Determinants of farm family income inequality. Am. J. Agric. Econ. 51:753-769.

Large, M. J. (1973) The Federal Environmental Pesticide Control Act of 1972, a compromise approach. Ecology L.Q. 3:277-292 (University of California School of Law, Berkeley).

McLendon, C. A. (1938) One variety community plan of cotton production shows many advantages and benefits for farmers. Ga. Agric. Exp. Stn. Mimeogr.

Norgaard, R. B., and A. Levinson (1974) An evaluation of integrated pest management programs for California and Arizona cotton. Prepared for CEQ-EPA.

Rodgers, W. H., Jr. (1970) The persistent problems of persistent pesticides: a lesson in environmental law. Columbia L. Rev. 1970:567-711.

Starbird, I. R., and B. L. French (1972) Costs of producing upland cotton in the United States, 1969. U.S. Dep. Agric. Econ. Res. Serv. Agric. Econ. Rep.

Street, J. H. (1957) The new revolution in the cotton economy. University of North Carolina Press, Chapel Hill, Chapter 11.

U.S. Department of Agriculture (1973a) Secretary's Memorandum No. 1799.

U.S. Department of Agriculture (1973b) Cotton varieties planted, 1969-1973. U.S. Dep. Agric. Marketing Serv.

Ware, J. O. (1938) Plant breeding cotton and cotton improvement. Farmers Short Course, Baton Rouge, Louisiana.

3

CONTEMPORARY PRACTICES FOR PEST MANAGEMENT

PLANT PATHOGENS

Pathogenic microorganisms attack all parts of the cotton plant. The diseases caused by these organisms may result in stunting or death of the plant, reduced or delayed fruiting, and fiber or seed deterioration. The direct costs to the cotton farmer include reduced yields, control measure expenditures, and reduced fiber and seed quality. Estimates of losses due to cotton diseases have been prepared annually for more than 20 years by the Cotton Disease Council from reports by extension and research personnel in the cotton producing states. Production loss due to the presence of the disease is estimated and expressed as a percent of actual yields. Loss estimates in recent years have ranged from 14 to 15 percent of annual production. Disease loss estimates by state and type of disease for 1973 are shown on Table 3-1.

Although disease losses occur in all cotton producing states, the types and severity of diseases differ regionally. Environmental factors such as humidity and temperature have a major influence on plant disease outbreaks. The range of control measures used to prevent or reduce disease losses includes cultural practices, use of disease resistant varieties, chemical seed treatment, chemical soil applications, and fumigation.

Current Practices

Verticillium wilt is a vascular disease caused by the microsclerotial (dark resting bodies, which are resistant to unfavorable conditions and may remain dormant for long periods of time) form of the fungus *Verticillium dahliae* (also designated *V. alboatrum*). This disease is present

TABLE 3-1 Estimated Reduction in 1973 Cotton Yield as a Result of Parasitic Disease Damage

Diseases	Ala.	Ariz.	Ark.	Calif.	Ga.	La.	Miss.	Mo.	N.M.	N.C.	Okla.	S.C.	Tenn.	Texas	Bales Lost	Percent Loss
Fusarium wilt	2.00	0	0.50	Tr.	1.50	5.00	0.10	1.00	0	0.50	2.25	1.50	0.25	0.60	95,637	0.73
Verticillium wilt	Tr.	1.49	1.00	5.30	Tr.	0	2.00	3.00	5.00	0.50	4.00	Tr.	1.00	4.00	375,088	2.87
Bacterial blight	Tr.	0	0.50	0	Tr.	3.00	1.00	7.00	Tr.	Tr.	2.25	0.30	0.50[d]	1.50	136,455	1.05
Phymatotrichum root rot	0	1.95	Tr.	0	0	0	0	0	0.10	0	Tr.	0	0	1.50	83,315	0.64
Seedling diseases	5.00	0.29	3.00	2.50	2.00	8.00	3.00	6.00	2.50	4.00	4.25	4.00	5.00	1.50	351,181	2.70
Ascochyta blight	Tr.	0	Tr.	0	Tr.	Tr.	0.01	Tr.	0	1.50	Tr.	Tr.	Tr.	0.40	21,488	0.16
Boll rots	5.00	1.63	2.00	1.20	1.50	18.00	4.50	3.00	Tr.	1.50	4.00	5.00	2.00	0.40	328,856	2.52
Nematodes	10.00	0.30	1.00	2.50	4.25	7.00	1.00	0.50	0.30	4.00	3.25	6.00	0.25	3.50	381,826	2.93
Others	0	0.13[a]	Tr.[b]	0	0	2.00[b]	0	0.50[b]	Tr.[a,b,c]	Tr.	2.00[b]	0.50	3.00[b]	0.10	40,893	0.31
Total Loss (%)	22.00	5.79	8.00	11.50	9.25	43.00	11.61	21.00	7.90	12.00	22.00	17.30	12.00	13.50	13.91	13.91
Total Bales Lost (thousands)	101.2	36.361	82.4	204.723	36.075	236.5	208.98	39.9	12.956	19.2	92.4	51.9	54	638.145	1,814.74	
Yield (thousand bales)[e]	460	628	1,030	1,780.2	390	550	1,800	190	164	160	420	300	450	4,727 =	13,049.2	

[a]Rust (*Puccinia cacabata*).
[b]Leafspots (*Alternaria*, *Cercospora*, *Phomopsis*, other species).
[c]*Thielaviopsis basicola* and other root rots.
[d]Loss to boll rot phase of this disease included under Boll Rots.
[e]Estimated yields from reporting states only, according to Statistical Reporting Service, USDA, Athens, Georgia, December 1, 1973, U.S. beltwide yield = 13,067.200.

SOURCE: 1973 Cotton disease loss estimate committee report, compiled by Johnny L. Crawford, Chairman.

throughout the Cotton Belt but is more prevalent in the irrigated areas of the Southwest. Resistance genetically developed in commercial cotton varieties is the primary control measure. The fungus persists for several years either in plant material or as microsclerotia in the soil. Thus, control by crop rotation is not always effective. A systemic fungicide (benomyl) placed in the soil before planting, appears capable of reducing losses due to Verticillium wilt. However, the quantities currently required to achieve control seem to preclude economical use of chemical control techniques. Soil fumigation with chloropicrin and methylbromide has been attempted. However, in California, cotton growth following fumigation has been excessively vegetative and minor element deficiency symptoms have been observed. Telone has shown some experimental promise as a soil fumigant with little subsequent deleterious effect.

Another vascular disease, Fusarium wilt, occurs primarily in the Southeast, Texas, and Oklahoma. The host specific fungus, *Fusarium oxysporum vasinfectum*, is the causal agent. The fungus attack is usually most pathogenic in sandy soils where the root-knot nematode *Meloidogyne incognita* is also a pest. At present, use of wilt and root-knot nematode resistant varieties provides the most satisfactory control. Since disease caused by the knot nematode apparently predisposes cotton to Fusarium wilt, control of nematodes by fumigation is often a major step toward controlling wilt. Crop rotation to control wilt is not entirely successful because *Fusarium* is capable of living dormantly as chlamydospores in the soil for several years. However, the root-knot nematode is an obligate parasite and, consequently, is readily controlled by crop rotation.

Cotton root rot is caused by the fungus *Phymatotrichum omnivorum* and is prevalent in the highly calcareous soils of the Southwest. The fungus, apparently native to these cotton growing areas, persists in the soil for many years and is capable of attacking any broadleaf host plant during the summer months. Because the sclerotia are produced on infected cotton roots at great depths, the disease usually occurs perennially in infested areas. Although early maturing varieties may partially escape effects of the disease, no resistant varieties have been developed. Incorporation of leguminous crops into the soil may reduce the severity of the disease. Although Benlate, a systemic fungicide, has been experimentally effective, no practical chemical control measures are currently available.

Bacterial blight, caused by *Xanthomonas malvacearum*, can attack any part of the plant aboveground at any stage of growth. It occurs mainly in the southeastern United States and Texas. Control is most often obtained by use of resistant varieties.

A large number of parasitic organisms, such as *Rhizopas, Nigrospurm*, and *Aspergillus*, may cause boll rots. One of the potentially most important is *Aspergillus flanus* which causes yellow stain of lint and aflatoxin in seed. This disease is limited to the lower deserts of the Southwest. The fungus penetrates bolls through exit holes made by the pink bollworm. Parasitic organisms may be classed as primary invaders capable of penetrating an uninjured boll, or as secondary invaders which must enter the boll through previously made wounds or openings. Areas with high rainfall and humidity, such as the Mississippi Delta, favor the development of boll rots. Some degree of control may be obtained by measures that open the leaf canopy to greater air circulation. These may include skip-row planting, limiting vegetative growth, planting varieties with smaller leaves (okra leaf character), and use of more open areas.

Seedling diseases of cotton are caused by seedborne or soilborne fungi and bacteria. Seedborne diseases may be controlled by chemical seed treatment, if the pathogen lies on the external surface of the seed coat. However, the pathogens causing anthracnose *(Glomerella gossypii)*, ascochyta blight *(Ascochyta gossypii)*, and bacterial blight *Xanthomanas malvacearum)* may be present under, as well as on, the seed coat. Thus, with currently available chemicals, complete control is difficult to achieve without harming the seed. *Rhizoctomia solani, Pythium ultimum*, and *Thielaviopsis basicola* are soilborne pathogens that usually damage plants in the juvenile stage. Seed treatment with fungicidal chemicals provides effective control only in the soil immediately surrounding the seed. An additional control measure is the application of fungicides in the row at planting. Treatment of the band of soil above the seed protects the emerging seedling. Seedling diseases are more damaging to cotton during unusually cool periods when emergence of the seedling is delayed. A rapidly growing seedling in a warm, dry climate may escape the disease. Although seed treatment may be necessary for many growing seasons, the unpredictability of weather dictates that nearly all cottonseed be treated with fungicides prior to sale for planting.

The root-knot nematode *(Meloidogyne incognita)* is the most important nematode pest of cotton. The female nematode attacks cotton roots causing hypertrophy of cells. This nematode is often a problem where cotton is grown on sandy soils. Preplant soil fumigation provides the most effective control; other less effective measures include fallowing, flooding (to decrease aeration), or rotating to nonsusceptible crops. Plant varieties resistant to root-knot nematodes exist; however, since those varieties show a strong positive yield response following soil fumigation, resistance is not totally effective.

The USDA conducted surveys on farm use of pesticides in 1964, 1966, and 1971. Table 3-2 shows the level of chemical pesticide use to control plant diseases and nematodes on cotton farms. Because the pesticide data are based on farm use, commercial seed treatment to control cotton diseases is not included. The data seems to reflect considerable variation in the quantities of fungicides and fumigants used during the survey years.* There is no apparent trend in the farm use of these chemicals. The data on acres treated must be viewed with caution, because any given acre may have been counted more than once, if it received more than one treatment. However, the data do indicate that the acreage treated with chemicals by farmers for cotton disease control is small relative to total cotton acreage.

WEEDS

Weed control in cotton presents complex problems. Because cotton is grown under different environmental conditions in different areas, weed species vary. Target weed species may be annuals, biennials, or perennials. These species include a wide range of plant types, from the most simple to the most complex plant forms and they vary widely in rooting depth, height, and spreading habits. Weeds growing in cotton can interfere with production by competing for available light, moisture, and nutrients. In some cases, allelopathic weed/crop interaction may inhibit cotton growth, thereby reducing cotton yields. Weeds also may increase the costs of production and reduce the quality and marketability of lint. For example, grasses can irreversibly stain the lint during harvest and leave plant fragments entangled with the gin lint. Such fragments are almost impossible to remove in spinning and are deleterious

---

*Some states collect data on the use of pesticides on cotton. For example, California produces an annual *Pesticide Use Report* containing detailed information on the use of fungicides and other pesticides. However, at present, it includes all chemicals applied by commercial pest control operators, but only some of the restricted chemicals applied by farmers.

TABLE 3-2 Quantities of Selected Fungicides and Fumigants Used on Cotton and Acres of Cotton Treated with Fungicides and Fumigants, United States, 1964, 1966, 1971

Chemical	Active Ingredients (1,000 lb)			Acres Treated[a]		
	1964	1966	1971	1964	1966	1971
Fungicides						
Inorganic fungicides						
Sulfur	10,245	1,027	15,078	738	155	667
Copper sulfates	--	--	26	--	--	52
Other coppers	7	13	--	46	23	--
Other inorganics	--	--	*b*	--	--	42
Organic fungicides						
Dithiocarbamates	*b*	15	46	*b*	15	29
Zineb	--	--	12	--	--	6
Ferbam	--	15	--	--	15	--
Others	--	--	34	--	--	23
Phthalimides	62	272	5	258	87	9
Captan	--	5	5	--	21	9
Others	--	267	--	--	66	--
Phenols	6	36	74	50	14	73
Pentachlorophenol	6	--	--	50	--	--
Karathane, dodine, quinones	*b*	--	15	*b*	--	27
Other organic	93	40	54	101	54	140
Fumigants	107	7,698	1,164	126	214	269
D-D mixture	*b*	3,789	--	*b*	49	--
Dibromochloropropane	--	--	211	--	--	24

TABLE 3-2 (Continued)

Chemical	Active Ingredients (1,000 lb)			Acres Treated[a]		
	1964	1966	1971	1964	1966	1971
Telone	--	--	616	--	--	14
Nemagon	--	3,197	--	--	67	--
Sulfur dioxide	--	564	--	--	33	--
Other fumigants	107	148	337	126	165	231

[a]The acres treated with a single chemical may be overstated if different commercial pesticides containing the same ingredients are applied to the same acreage in separate treatments. Also, the number of acres treated with different ingredients should not be added together because two or more of these ingredients may have been used separately or together on the same acreage.
[b]Less than 500 lb or less than 10,000 acres.

SOURCE: *Quantities of Pesticides Used by Farmers in 1964*, Agricultural Economic Report #131, ERS, USDA; *Quantities of Pesticides Used by Farmers in 1966*, Agricultural Economic Report #179, ERS, USDA; *Quantities of Pesticides Used by Farmers in 1971*, (mimeo).

to dye and other finishing processes. Weeds also harbor plant pathogens, insects, and rodents.

Cotton losses due to the presence of weeds may be severe although the damage caused is not always as obvious as losses caused by other pests. The USDA has estimated that annual losses in the form of reduced yields and quality caused by weeds averaged about eight percent of the value of the cotton crop during the years 1951-1960 (USDA 1965).

Weed control is important through the entire growing season. The first plant to occupy an area has an advantage over later arrivals. Thus, early control of weeds is directed at establishing the crop plants. Late season weed control is important in maintaining the quality and quantity of harvested cotton. (One mature large crabgrass plant per six meters of cotton row is sufficient to reduce the grade of mechanically harvested cotton.)

Current Practices

In contrast to other types of pest control activities, no method of weed control has ever been completely discarded. Older methods still have their place and are often combined with new procedures. Unfortunately, the weed species that may infest any given area are so numerous and diverse that any single control method is rarely satisfactory.

Weed control techniques may include crop manipulation or habitat management. The objectives are to reduce or eliminate competition from weeds and to prevent their introduction or spread. A rotation of crop species involves varying soil manipulation and planting and harvesting procedures. Such variations may make it difficult for a single weed species to become a major problem in a specific area. However, crop rotation is not widely used in the cotton growing areas of the United States.

Physical control methods are probably the oldest procedures for controlling weeds. Although physical methods are still used, these methods alone cannot provide the degree of control needed for modern agricultural production. Variable climatic conditions, weed seed dormancy, presence of various perennial underground storage organs, production of thousands of seeds per weed plant, continual introduction of seed, and wide dissemination of seed by man and nature combine to prevent control of many species by physical methods. However, several of these methods are still widely practiced, including tillage (physically disturbing weeds in the soil), pulling or hoeing, flooding, draining, smothering, and burning. Of these procedures, tillage is the most widely used.

Tillage for weed control in cotton includes all soil disturbing procedures, including mechanical land preparation and machine cultivation. Preplanting tillage (seedbed preparation) removes germinated weed seedlings but does not significantly deplete seed reserves in the soil. Postplanting tillage (cultivation) removes or buries weeds, but weakens cotton plants through root pruning or other injuries. (Under a majority of soil conditions, cultivation is of little benefit other than for weed control.) Generally, tillage operations are effective against seedling weeds, but are relatively ineffective against intact seeds in the soil, large weeds, or perennial weed species. In most of the Cotton Belt, cultivation is the major cultural weed control practice. The actual number of cultivations may vary from one to six per season. Cultivation can be used successfully between the rows but is much less effective in controlling weeds within the row. Adverse weather conditions may severely limit the usefulness of many tillage operations. Also, the continued operation of heavy tillage implements over a field has been shown to be very detrimental to soil structure.

Another long-established cultural control practice is the use of either "flamers" or herbicidal oils which burn the weeds. These control only very small weeds and require careful timing and precise assessment of soil and crop conditions.

Extensive use of herbicides to control weeds in cotton began in the 1950s. At present, most U.S. cotton acreage receives one or more herbicide treatments annually. In 1968, approximately 90 percent of the total cotton acreage was treated with herbicides, at an average cost of $9.66 per acre (USDA 1972). Two factors appear to have significantly influenced the use of herbicides. One factor has been the relative cost of farm labor and herbicides for weed control. Herbicide application has been substituted for more expensive hand labor. A second factor has been government farm program acreage control and price support criteria. These provide strong incentives for increasing per acre yields by more intensive use of inputs such as herbicides.

Herbicides may be applied during several phases of the crop year. Preplant treatments may be applied from three months prior to planting to the day of planting. Preemergence herbicides are applied to planting or a few days thereafter. Postemergence treatments are applied after emergence of the cotton plant, usually during the first three to five weeks of plant growth. However, applications may continue throughout the growing season to control weeds which may interfere with harvest. In 1968, approximately

3.4 million cotton acres received only preplant or preemergence treatments at an average cost of $6.54 per acre. Postemergence treatments only were applied to 1.2 million acres at an average cost of $4.52 per acre. Approximately 4.6 million acres received both preplant or preemergence and postemergence herbicides, at an average cost per acre of $13.32.

The heaviest use of herbicides in cotton production occurs in the Mississippi Delta. Over 90 percent of the acreage in this area was treated with a preplant or preemergence herbicide. About one-third of the acreage receives a "dual" or "overlay" treatment--application of a preemergence herbicide after planting over a preplant herbicide treatment. The primary reason for "dual" treatments is to combine the preplant herbicide effectiveness for grass control with the preemergence herbicide effectiveness for broadleaf weed control.

Most preplant herbicides used on cotton are members of the dinitroaniline herbicide family. Substituted-urea herbicides are sometimes used for either preplant or preemergence treatments. In the western half of the Cotton Belt, members of the *s*-triazine family of herbicides are frequently used. These three chemical families plus the organic arsenicals comprise most of the herbicides used in cotton production. (The more widely used herbicides in these groups are: trifluralin in the dinitroaniline group; fluometuron in the substituted-urea group; prometryn in the *s*-triazine group; and MSMA in the organic arsenical group.) Table 3-3 shows herbicides used on cotton in 1964, 1966, and 1971.

Postemergence herbicide treatments do not reduce weed competition as effectively as preplant or preemergence treatments. They are most efficient when applied to small and actively growing weeds. Organic arsenicals are the most widely used postemergence herbicides. Their major use is for control of Johnson grass, nutsedge, and annual grasses. (See Table 3-4 for scientific names of weeds.) Depending on the weed species to be controlled, many other herbicides can be used postemergence. The level of postemergence herbicide usage varies among the major cotton growing areas. Growers in the Mississippi Delta, Lower Rio Grande Valley, and California tend to apply postemergence herbicides more frequently than growers in other areas.

As chemical weed control procedures have become more selective with less potential injury to the crop, the varieties of problem weeds in cotton are changing. Annual grass weed species are declining in importance, primarily because of the effectiveness of dinitroaniline herbicides.

TABLE 3-3 Quantities of Selected Herbicides Used on Cotton and Acres of Cotton Treated with Herbicides, United States, 1964, 1966, 1971

Type of Herbicide Product	Active Ingredients (1,000 lb)			Acres Treated (1,000)[a]		
	1964	1966	1971	1964	1966	1971
Inorganic Herbicides[b]	1,393	1,458	557	94	542	265
Organic Herbicides						
Arsenicals[b]	986	802	7,569	578	1,020	4,123
Phenoxy[c]	202	267	65	225	182	296
2,4-D	202	250	4	225	175	5
2,4,5-T	--	[d]	--	--	3	--
MCPA	--	17	[d]	--	4	4
Other	--	--	61	--	--	287
Phenol Urea	1,094	1,180	3,997	3,157	1,559	5,260
Monuron	80	--	--	110	--	--
Diuron	991	882	568	2,996	1,282	776
Linuron	--	6	53	--	9	220
Fluometuron	--	--	3,334	--	--	4,206
Other	23	292	42	51	268	58
Amides	--	--	196	--	--	--
Alanap	--	--	4	--	--	17
Alachlor	--	--	4	--	--	2
Other	--	--	188	--	--	108
Carbamates	214	6	3	102	14	18
CIPC & IPC	208	1	--	90	6	--
Other	6	5	3	12	8	18
Dinitro Group	[d]	11	382	[d]	13	256
Triazines	19	--	806	23	1	1,118
Atrazine	19	--	--	23	--	--

TABLE 3-3 (Continued)

Type of Herbicide Product	Active Ingredients (1,000 lb)			Acres Treated (1,000)[a]		
	1964	1966	1971	1964	1966	1971
Propazine	--	[d]	--	--	1	--
Simazine	--	--	--	--	--	--
Other	[d]	--	806	[d]	--	1,118
Benzoic	--	--	--	--	--	--
Amiben	--	--	--	--	--	--
Chlorinated aliphatic	35	--	--	35	--	--
Trifluralin	519	2,631	4,544	813	3,552	6,804
Nitralin	--	--	500	--	--	516
Amitrole	--	--	--	--	--	--
Dalapon	--	--	18	--	--	22
Norea	--	--	846	--	--	419
Other organic	146	171	127	180	283	419
TOTAL	4,628	6,526	19,610			

[a]The acres treated with a single herbicide may be overstated if different commercial pesticides containing the same ingredients are applied to the same acreage in separate treatments. Also, the number of acres treated with different ingredients should not be added together, because two or more of these ingredients may have been used separately or together on the same acre.
[b]May include some materials that were used as defoliants or desiccants.
[c]Used primarily as desiccants.
[d]Less than 500 pounds or less than 10,000 acres treated.

SOURCE: *Quantities of Pesticides Used by Farmers in 1964*, Agricultural Economic Report #131, ERS, USDA; *Quantities of Pesticides Used by Farmers in 1966*, Agricultural Economic Report #179, ERS, USDA; *Quantities of Pesticides Used by Farmers in 1971* (mimeo).

TABLE 3-4 Major Weeds Troublesome in Cotton in the United States[a]

Weed		Regions			
Common Name	Scientific Name	South-east	Mid-South	High Plains[a]	West
Annual morning glory	*Ipomoea* spp.	X	X	X	X
Annual sedges	*Carex* spp.	X	X		
Barnyard grass	*Echinochloa* spp.	X	X		X
Bermuda grass	*Cynodon dactylon* (L.) Pers.	X	X		
Broadleaf signalgrass	*Brachiaria platyphylla* (Griseb.) Nash		X		
Carpetweed	*Mollugo verticillata* L.	X			
Cocklebur	*Xanthium* spp.	X	X		
Common lamb's-quarters	*Chenopodium album* L.	X		X	X
Common purslane	*Portulaca oleracea* L.	X	X	X	X
Crabgrass	*Digitaria* spp.	X	X		X
Crowfoot grass	*Dactyloctenium aegyptium* (L.) Richter	X	X		
Florida purslane	*Richardia scabra* L.	X	X		
Goose grass	*Eleusine indica* (L.) Gaertn.	X	X		
Groundcherry	*Physalis* spp.			X	X
Horse nettle	*Solanum* spp.	X	X	X	X
Jimsonweed	*Datura stramonium* L.	X			
Johnson grass	*Sorghum halepense* (L.) Pers.	X	X	X	X
Nutsedge	*Cyperus* spp.	X	X		X
Pigweed	*Amaranthus* spp.	X	X	X	X
Poorjoe	*Diodia teres* Walt.	X			
Puncture vine	*Tribulus terrestris* L.			X	X
Ragweed	*Ambrosia* spp.	X			
Redvine	*Brunnichia cirrhosa* Gaertn.		X		
Sicklepod	*Cassia obtusifolia* L.	X			
Sida	*Sida* spp.	X	X		
Sprangletop	*Leptochloa* spp.	X	X		X
Witchgrass	*Panicum capillare* L.	X	X		X

[a]Texas and Oklahoma.

SOURCE: Adapted by permission from J. T. Holstun, Jr., and O. B. Wooten, "Weeds and Their Control," in Fred C. Elliot, Marvin Hoover, and Walter K. Porter, Jr., eds., *Advances in Production and Utilization of Quality Cotton*, © 1968 by Iowa State University Press, Ames, Iowa.

Although annual broadleaf weeds are becoming less prevalent, some specific annual broadleaf species not susceptible to currently used herbicides are becoming more prevalent. Examples include prickly sida, spurred anoda, velvetleaf, spotted spurge, cocklebur, and several morning glory species. Some of these species are in the same taxonomic family as cotton and exhibit similar herbicide resistance characteristics.

Although a portion of the prevalent weed spectrum is changing to nonsusceptible annual broadleaf species, the major change has been the encroachment of several perennial weed species into cotton fields. All of the selective herbicides currently used for preplant or preemergence treatments in cotton kill weeds by affecting germinating seedlings before they reach one to two inches in height. However, Johnson grass and nutsedge often grow from vegetative reproductive organs in the soil (rhizomes or nutlets). Plants developing from these organs are resistant to the action of most selective cotton herbicides. In addition, several perennial broadleaf weed species are becoming problems. Silverleaf nightshade and Carolina horse nettle in the West, and honeyvine milkweed and redvine in the Southeast are not adequately controlled by current practices.

Several trends in herbicide usage seem likely to continue. Growers are beginning to use herbicide combinations to broaden the spectrum of weed control. In addition, rotation of herbicides may be more widely used. Such rotation would require a larger herbicide arsenal. Continued use of a single herbicide or biological control system can result in a fairly rapid change in weed composition. Rotation of herbicides is seen as a technique for reducing such change. The increasing problem of controlling perennial weed species has increased rates of herbicide applications. (Growers have attempted using nonchemical control measures, such as additional cultivation and disking. Unfortunately, these measures often spread vegetive reproductive organs throughout the field.)

ARTHROPOD PESTS

In every cotton producing area of the United States, there is a major pest insect which must be controlled if the crop is to be profitably produced. The single most important cotton pest is the boll weevil, *Anthonomus grandis* (Boheman), a devastating species that invaded the United States from Mexico in the 1890s. A second invader is the pink bollworm, *Pectinophora gossypiella* Saunders, which also entered the United States from Mexico. This pest has been particularly destructive to cotton grown in the irrigated deserts of the western United States. More recently, two species of bollworms, *Heliothis zea* Boddie and *H. virescens* Fabricus, have inflicted great damage to cotton crops in all production areas. Certain hemipterous plant bugs, particularly the cotton fleahopper, *Pseudotomoscelis seriatus* (Reut.) and *Lygus* spp., are also serious pests. The bollworms and hemipterans apparently are indigenous to North America. Cotton also is attacked by many secondary and occasional pests including mites, thrips, aphids, and several leaf-feeding caterpillars. As shown in Table 3-5, these pests caused losses to the cotton crop of almost $500 million each year (1951-1960).

TABLE 3-5 Estimated Average Annual Losses to Cotton Caused by Insects and Spider Mites, 1951-1960[a]

	Loss from Potential Production		
Pest	Percent	Quantity[a] (1,000 bales)	Value[b] ($1,000)
Boll weevil	8.0	1,239	200,613
Bollworm and tobacco budworm	4.0	619	100,307
Plant bugs	3.4	527	85,261
Spider mites, pink bollworm, cotton aphid, cotton leafworm, thrips, cotton leaf-perforator, cabbage looper, and beet armyworm	3.6	558	90,276
TOTAL	19.0	2,943	476,457

[a]Estimates based on full production with causes of loss eliminated.
[b]Includes quality and quantity and assumes that market outlets would be available for increased production with no change in average prices.

SOURCE: USDA 1965.

The boll weevil is a serious pest from the Atlantic coast to the southern high plains of Texas. The pink bollworm has not achieved pest status in the humid (rainbelt) portions of the Cotton Belt except in Arkansas and Louisiana. However, it is widespread in Texas, Oklahoma, New Mexico, Arizona, and California. In most areas, the pink bollworm has been adequately controlled by cultural practices; in Arizona and southern California it must be controlled with insecticides. The *H. zea* bollworm is a pest problem throughout the Cotton Belt. While the tobacco budworm, *H. virescens*, is restricted to the rainbelt and Texas, it causes serious damage within these regions. Lygus bugs cause varying degrees of damage over a wide area, but are most destructive in the San Joaquin Valley of California and the Mississippi Delta. The cotton fleahopper is a major pest in Texas and Oklahoma.

Spider mites, *Tetranychus* spp., may be pests in any of the cotton growing areas of the United States, but are particularly apt to appear in the desert regions. Spider mite outbreaks are generally more localized and of less serious consequence than those of the pest insect species.

The abundance of each pest and the amount of damage inflicted upon crops by pests are greatly influenced by climate and diversity of the ecosystems in which cotton is produced. In broad terms, the ecosystems and vulnerability to insect pest species may be described as follows:

1. In the *irrigated deserts* of the Far West the major pests are the pink bollworm, lygus bug, and lygus bollworm.
2. In the *semiarid regions* of the southwestern United States the bollweevil, fleahopper, bollworm, and tobacco budworm are the major pests.
3. In the *humid regions* of the mid-South and southeastern United States the boll weevil, plant bug, bollworm, and tobacco budworm are the major pests.

Climate, especially the amount of rainfall, in each of these regions differs. Also, in each of these regions, cotton is grown under different production schemes and in combination with different crops. Cotton and alfalfa are grown in California, cotton and grain sorghum in Texas and Oklahoma, and cotton and soybeans in the South. These regions represent different kinds of agroecosystems and the type of pest management practiced in each has unique features.

Briefly, cotton insect pest control in each of these regions may be summarized as follows: in the irrigated West, *Lygus* spp. and the pink bollworm are of primary importance and control is usually achieved through the use of insecticides. These treatments often eliminate the arthropod natural enemies which suppress the bollworm, cabbage looper [*Trichoplusia ni* (Hubner)], the beet armyworm [*Spodoptera exigua* (Hubner)], spider mites, and other secondary pests. These pests, freed from their natural enemies, frequently increase in abundance and cause severe losses. Chemical control of these pests is also difficult. Exploding pest populations often require repeated and costly insecticidal treatments. Thus, directly and indirectly, lygus and pink bollworms instigate insecticidal treatment of thousands of acres of cotton in California, Arizona, New Mexico, and west Texas (Falcon et al. 1968, Smith and Falcon 1973, Stern 1969, van den Bosch et al. 1971, Adkisson 1973).

Insecticidal treatment of cotton offers the only means for controlling damaging outbreaks of the boll weevil and cotton fleahopper in the semiarid regions of Texas and Oklahoma. Unfortunately, these treatments also eliminate the arthropod natural enemies of the bollworm and tobacco budworm. Each year tremendous outbreaks of these two pests occur on vast acreages of cotton. Control of these secondary pest outbreaks to prevent severe losses to crops then requires extensive use of insecticides for the remainder of the season. Crop failures have resulted in some areas because tobacco budworm has developed resistance to most insecticides. In areas where insecticidal control of the boll weevil and fleahopper may be avoided in early and mid-season, widespread outbreaks of the bollworm and budworm seldom occur and the intensive use of insecticides on cotton is averted. Pest control programs in these regions center on cultural practices directed toward suppression of boll weevil and pink bollworm populations combined with the selective use of minimal amounts of insecticides against the fleahopper and boll weevil. The strategy is to keep the two key pests below crop-damaging numbers by methods which conserve insect natural enemies and do not induce bollworm-budworm outbreaks (Adkisson 1972, Sterling and Haney 1973).

Insect pest control in cotton in the humid South and southeastern United States is much the same as in Texas and Oklahoma, except that the plant bug *L. lineolaris* replaces the fleahopper as a major pest during the early growing season. The boll weevil is perennially the most damaging pest, but bollworm and tobacco budworm outbreaks are becoming increasingly serious. Although outbreaks of the bollworm and budworm may occur in localized areas in the absence of insecticidal treatments of cotton for control of other pests, most widespread outbreaks are induced by early insecticidal treatment to control the boll weevil or plant bug, or by premature applications against the bollworm. Spider mites occasionally become serious, especially under drought conditions or following insecticidal treatment of cotton for other pests (Newsom and Brazzel 1968, Newsom 1972).

Given our present state of knowledge, one or more major pest insects in each cotton production area in the United States, must be controlled with insecticides if cotton is to be grown profitably. However, the very use of these chemicals often creates problems more serious than those which they were applied to ameliorate; these problems include target pest resurgence, secondary pest outbreaks, and pest resistance to insecticides. This costly insecticidal treadmill in which initial insecticidal treatments create need for continued treatment is driving some cotton producers to economic ruin and is also contributing significantly to environmental pollution (Smith 1971; Adkisson 1969).

Evolution of Control Measures

Cotton has serious insect problems in every producing region. Perhaps because of these problems and the crop's economic importance, more pest control research has been conducted on cotton than on any other commodity. This research has led to intensive crop protection measures designed to produce maximum yields. Patterns of crop protection developed during the history of cotton production have evolved through a series of phases from the most primitive to the most sophisticated. Although all phases are presently used somewhere in the world today, most U.S. cotton protection is in the final three phases. This evolution of pest control on cotton has been described as follows:

Subsistence Phase

Cotton is usually grown under nonirrigated conditions as a part of a subsistence agriculture. The cotton is usually processed by native hand weavers and does not enter world trade. Yields are low (less than 200 pounds of lint per acre). There is no organized program of crop protection. Whatever crop protection occurs is the result of natural control, pest resistance inherent in native varieties, hand picking of pests, cultural practices, and rare pesticidal treatments.

Exploitation Phase

Irrigated lands are opened for crop production. Cotton is one of the first crops planted to exploit this resource. Alternatively, cotton is moved to fertile land and planted in large acreages under the plantation system. Crop protection measures are introduced to protect the larger and more valuable crop. New higher yielding varieties (which may be less pest resistant) are introduced and crop protection methods are intensified. In most cases insect pest control relies solely on chemical insecticide treatment used intensively and often on fixed schedules. These methods are successful initially and high yields are obtained.

Crisis Phase

After a varying number of years in the exploitation phase characterized by heavy use of chemical pesticides, a number of important changes occur. More frequent treatment of the crop becomes necessary to achieve effective pest control. The treatments must be started earlier in the growing season, made at more frequent intervals, and extended later into the harvest season. Pest populations now resurge rapidly after treatment to new and higher levels. Some pests become so resistant to specific pesticides that application becomes useless. As other pesticides are substituted, the pest populations develop resistance to them too; unfortunately, this happens more rapidly than with the first chemical. Simultaneously, secondary or occasional pests, which have never been a serious problem previously, become serious and regular ravagers of cotton fields. Thus, the combination of pesticide resistance, target pest resurgence, and the unleashing of secondary pests or the induced rise of a previously innocuous species to pest status, results in greatly increased pesticide use which greatly increases production costs.

Disaster Phase

Heavy pesticide usage combined with increased crop losses inflicted by uncontrolled pests makes the production of cotton unprofitable. At first only marginal production is eliminated, but eventually all cotton production may cease in the affected area. This has occurred in several Central American countries, parts of Mexico, and in South Texas. However, in most areas of the United States the disaster phase has been postponed by implementing advanced pest control technology and a government subsidy program which has provided some economic advantage to cotton producers.

Integrated Control and Recovery Phase

The integrated control phase has developed in only a few places. The most noteworthy examples are in several valleys in Peru and the Lower Rio Grande Valley of Texas (Smith and Reynolds 1972, Adkisson 1972). In the integrated control phase, the crop protection system relies on a variety of control procedures rather than pesticides alone. Attempts are made to manipulate environmental conditions to suppress pest populations. Fullest use is made of natural biological control and other natural regulating factors. Minimal amounts of chemical pesticides are used only to suppress pest insect outbreaks which have attained crop damaging levels. Maximum use is made of cultural and biological control methods. Supervised control procedures are used to monitor insecticide application. This combination of procedures allows cotton to be produced profitably in the area again.

The evolution of crop protection patterns in cotton shows that pest controls cannot be developed to control any single pest. Instead, pest management measures must be developed and applied with reference to the particular ecosystem within which crops, production practices, and existing numbers of pests (or potential populations) exist and interact. New technologies developed for control or management of pest insects in the future must be applied with knowledge of their impact on the various interactions operating in the affected ecosystem.

Chemical Control

Chemical approaches to the control of arthropod pests of cotton may be divided into four periods: time prior to the general use of insecticides (pre-Civil War times until 1923); the period characterized by extensive use of calcium arsenate-nicotine sulfate dusts (1924-1945); the chlorinated hydrocarbon insecticide era (1945-1955); and the organophosphorus insecticide era after the boll weevil became resistant to the chlorinated hydrocarbons (1955 to present) (Newsom and Brazzel 1968).

Prior to the invasion of the United States by the boll weevil in 1892, there was relatively little damage to cotton by insects or spider mites. There were no major pests. The bollworm, cotton leafworm [*Alabama argillacea* (Hubner)], and the cotton aphid (*Aphis gossypii* Glover) infrequently appeared in sufficient numbers to cause alarm among growers. Most cotton growers did not attempt to control these pests with insecticides. A few used Paris green, London purple, lead, or calcium arsenate against the worms and nicotine sulfate against the aphids. This changed when the boll weevil entered the United States and inflicted great losses. At that time, an intensive search began for more effective insecticides and improved application machinery. The first insecticide recommended for control of the boll weevil was an arsenical molasses spray containing Paris green, London purple, or lead arsenate (Townsend 1895, Malley 1901). However, the level of control achieved was unsatisfactory because of poor formulations and inadequate machinery for applying the sprays.

The airplane introduced the era of chemical pest control to the cotton industry. In 1921, the Ohio Experiment Station first demonstrated the utility of aerial application by controlling a catalpa sphinx *Ceratomia catalpae* (Bois.) outbreak on catalpa trees. The first airplane experiments on cotton were conducted in 1922 at the USDA Federal Delta Laboratory by Coad and McNeil (1924). They obtained effective control of the cotton leafworm with aerial applications of calcium arsenate dusts. In 1923, the two entomologists demonstrated that the boll weevil also could be effectively controlled with aerial applications of calcium arsenate dusts. The success of this method was confirmed in Georgia and Texas during the period 1925-1927 (Post 1924, Thomas et al. 1929). (Dusting cotton by airplane was the principal insecticide application method until the early 1950s when low-volume sprays were developed. Sprays have now replaced dusts and dusting machines are infrequently used.)

The cotton aphid became a serious pest of cotton when the widespread use of calcium arsenate destroyed its natural arthropod enemies. Nicotine sulfate was added to the arsenical dust whenever this pest became a problem. The bollworm also became much more serious during this period for the same reason. However, calcium arsenate was fairly effective against very young larval bollworms and provided satisfactory control when applied to the crop regularly at short intervals.

The importance of the cotton leafworm declined after the introduction of calcium arsenate dusts. This decline continued with the introduction of new synthetic insecticides after World War II and the species is no longer economically important in the United States.

The cotton fleahopper and lygus were recognized as pests in the late 1920s, principally in the Southwest and the new irrigated areas of the West. Sulphur dusts were the principal means used to control these pests, as well as to control occasional spider mite outbreaks.

In summary, the modern era of intensive insecticidal treatment of cotton began in 1923 with the advent of airplane dusting. The principal insecticides used from this time until the introduction of chlorinated hydrocarbon materials were dusts of calcium arsenate, nicotine sulfate, and sulphur. Most insecticide use was confined to boll-weevil-infested areas.

The era of intensive insecticide treatment began at the end of World War II with the introduction of DDT, BHC, and toxaphene. These were followed by other synthetic organic chlorinated hydrocarbon insecticides including aldrin, dieldrin, endrin, heptachlor, Strobane, and TDE. The new insecticides possessed two qualities of great importance: (1) high initial toxicity to the cotton pest insects, and (2) sufficient persistence to control newly emerging insects or insects migrating from untreated to treated areas.

The chlorinated hydrocarbon insecticides had a great impact on cotton production. For the first time, cotton producers were able to achieve highly effective control of all arthropod pests of the crop. The impact of these insecticides was to stimulate an unprecedented demand by growers for almost complete control of pest insects. It then became profitable for producers to use fertilizer, irrigation, and long-growing indeterminate varieties. Spectacular yield increases were obtained at high profit levels for many years (Table 3-6).

The apparent victory over pest insects of cotton through intensive use of the chlorinated hydrocarbon insecticides

TABLE 3-6 Comparative Efficacy of Inorganic and Synthetic Organic Insecticides for Control of Cotton Pests as Reflected by Yield[a]

Type of Insecticide and Location	Number of Plots		Average Number of Applications	Average Yield		Increase Due to Treatment (%)		
	Treated	Untreated		Pounds	Lint/Acre[b]	Minimum	Maximum	Average
Inorganic insecticides								
Central Texas, 1939-1945	299	287	5.3	383	294	9.9	108.0	34.3
Florence, South Carolina, 1928-1945	551	392	6.9	441	344	5.3	98.0	23.6
Madison Parish, Louisiana, 1929-1945[c]	--	--	--	--	--	1.8	54.7	22.3
Synthetic organic insecticides								
Central Texas, 1946-1958	1,086	500	5.0	375	251	25.5	156.6	53.6
Florence, South Carolina, 1946-1958	1,252	517	7.8	519	343	15.1	217.9	53.9
Madison Parish, Louisiana, 1946-1956[d]	--	--	--	--	--	13.7	112.8	41.7

[a]Data from Gaines 1959 (42), Parencia 1959 (68), and Fye et al. 1962 (38) as discussed in Newsom and Brazzel (1968).
[b]Converted from seed cotton calculated at 36 percent lint.
[c]Average yield for parish, 319 lb, 1928-1945.
[d]Average yield for parish, 435 lb, 1946-1956.

was not lasting. By the mid-1950s, the boll weevil in Louisiana and Mississippi had developed resistance to these chemicals. The resistant pest strain spread rapidly across the southern and southwestern states and all infested areas were reporting chlorinated-hydrocarbon-resistant weevils by 1960 (Roussel and Clover 1955, Brazzel 1961).

This problem was solved by a switch to the organophosphorus (OP) and carbamate insecticides, mainly methyl parathion, azinophosmethyl, ethyl parathion, malathion, EPN, and carbaryl. The organophosphorus compounds were highly toxic to the boll weevil at relatively low concentration rates. However, these rates were not sufficient to control the bollworm and tobacco budworm. In order to gain control of all the major pest species, mixtures of chemicals were formulated containing DDT, toxaphene-DDT (and to a lesser extent, endrin) for control of the bollworm and budworm. Methyl parathion, azinophosmethyl, EPN, or malathion was added for control of the boll weevil.

These broad spectrum formulations also controlled aphids, fleahoppers, plant bugs, spider mites, and leaf-feeding caterpillars. Of course, they were also highly effective against insect parasites and predators. But, during this period most cotton producers were demanding insecticidal mixtures capable of "sterilizing" a field, this is, rendering them almost completely devoid of insects (Adkisson 1969, Reynolds et al. 1975).

In the early 1960s, the bollworm and tobacco budworm developed high levels of resistance to the chlorinated hydrocarbon and carbamate insecticides (Brazzel 1963, Brazzel 1964, Adkisson 1969, Harris et al. 1972). Pest control priorities in cotton suddenly reversed. The bollworm and tobacco budworm were more important pests than the boll weevil. The problem of bollworm and tobacco budworm resistance was solved by increasing the dosage of methyl parathion from 0.25 to 0.50 pound rate per acre per application for boll weevil control to 1.0 to 2.0 pounds per acre. Monocrotophos at 0.8 to 1.0 pound also was introduced as were mixtures containing 2.0 pounds of toxaphene, 1.0 pound of DDT, and 0.5 to 1.0 pound of methyl parathion. An immediate effect of increasing chemical concentration rates was increased production costs; yields remained high, but profits decreased (Adkisson 1969). This situation prevailed until the late 1960s when the tobacco budworm in the Lower Rio Grande Valley of Texas and northeastern Mexico became resistant to the organophosphorus insecticides. Many Valley producers treated fields with methyl parathion 15 to 18 times per year, but still suffered great losses in yield. Others produced at relatively high levels, but made small profits because of high costs incurred by intensive insecticidal treatment. Some cotton crops were destroyed regardless of treatment and, in fact, approximately 700,000 acres in northeastern Mexico were removed from cotton production because of budworm losses (Adkisson 1969, 1972; Reynolds et al. 1975).

Organophosphate-resistant tobacco budworms now occur in Texas, Louisiana, Arkansas, and other states to the east. The pest has developed such a high level of resistance that control has become difficult with any insecticide presently registered for use on cotton.

Another drastic change in the pesticide usage pattern on cotton occurred when the EPA banned the use of DDT beginning in 1973. DDT combined with toxaphene had provided satisfactory control of the boll weevil, bollworm, cotton fleahopper, and plant bugs in cotton producing states east of Texas. (Methyl parathion was frequently added at a low rate if weevils became extremely numerous.) Cotton producers in the states east of Texas had not faced severe pest resistance problems because toxaphene-DDT formulations had controlled a broad spectrum of pests and because resistance had developed slowly in the bollworm-tobacco budworm populations. The banning of DDT has forced southern cotton producers to shift to high concentrations of OP insecticides for pest insect control. Usually, these are applied in combination with toxaphene and, to a lesser extent, with endrin or chlordimeform. Thus, the banning of DDT has increased selection pressure for the development of OP-resistant pest strains. In due course, the cotton producers of the South may suffer the chemical pest control problems of decreasing effectiveness and increasing cost induced by OP insecticides which now are confined to Texas and Mexico. That is, chemical control will become more difficult, more costly, and less effective (Reynolds et al. 1975).

The evolution of insecticide control of pest insects in cotton grown in the irrigated deserts of the western United States where the boll weevil does not occur has followed a similar course, except that the dominant pests have been lygus, the bollworm and, more recently, the pink bollworm. In the deserts, lygus has been the primary pest for which insecticide treatment has been initiated. However, treatments also have been initiated to control the bollworm. These treatments kill insect natural enemies, unleashing attacks of bollworms, cabbage loopers, cotton leaf perforators [*Bucculattrix thurberiella* (Busck)], salt marsh caterpillars [*Estigmene acraea* (Durry)], and spider mites. Thus, in the desert regions, as in the boll-weevil-infested areas, once insecticidal treatments are initiated, they must be continued to control target pest resurgence and secondary pest outbreaks. This has led to intensive insecticidal treatment of cotton in desert areas (van den Bosch et al. 1971).

Severe insecticide-induced problems with secondary pests led to development of an integrated pest management system in California during the late 1950s. It concentrated on conserving the insect natural enemies of target and secondary pest species. Broad spectrum chlorinated hydrocarbon insecticides were phased out and selective dosages of certain OP insecticides used under supervised conditions when pests were causing crop damage were substituted. Strip harvesting of alfalfa and interplanting with cotton held lygus in alfalfa (the preferred host) forestalling lygus invasion of cotton (Stern et al. 1964). All these methods reduced the need for insecticide applications (Reynolds et al. 1975). This type of integrated program in Arizona and southern California was destroyed by the pink bollworm invasion of the late 1960s which prompted intensive treatment with OP insecticides. There has been a great resurgence of secondary pests. Presently, the use of insecticides in these areas is as great as in the boll-weevil-infested regions of the Cotton Belt (Smith 1971).

Present Status of Control Measures

Most recent statistics indicate that 54 percent or 270 million pounds of the approximately 500 million pounds of pesticides applied each year to U.S. cropland are insecticides. However, only about 5 percent of the total crop acreage is treated with insecticides. Approximately 47 percent of the pesticides applied to crops is applied to cotton (Eichers et al. 1970, Pimentel 1973). Amounts of insecticide used on cotton vary widely across the United States. Almost half (46 percent) of the cotton acreage receives no treatment in most years. The highest percentage (79 percent) of acres treated is in the Southeast and Delta States; the lowest percentage (37 percent) is in the Southern Texas Plains and the Northern Delta of Arkansas, Missouri, Tennessee, Kentucky, and Illinois (Pimentel 1973). Available statistics on total amounts of insecticide show that 78.0, 64.9 and 73.4 million pounds, respectively, were applied to U.S. cotton in 1964, 1966, and 1971 (Table 3-7).

TABLE 3-7 Quantities of Selected Insecticides Used on Cotton and Acres of Cotton Treated with Insecticides, United States, 1964, 1966, 1971

Type of Insecticide Product	Pounds of Active Ingredients (x1,000)			Acres Treated (x1,000)[a]		
	1964	1966	1971	1964	1966	1971
Inorganic insecticides	2,518	--	69	57	--	23
Botanicals and biologicals	--	2	--	--	8	--
Synthetic organic insecticides						
Organochlorines	55,778	49,703	42,619	14,252	10,157	6,130
Lindane	540	163	--	636	298	--
Strobane	--	2,016	216	--	225	18
TDE (DDD)	191	167	--	61	33	--
DDT	23,588	19,213	13,158	6,901	4,767	2,383
Methoxychlor	--	6	--	--	6	--
Endrin	1,865	510	1,068	1,194	403	262
Heptachlor	[b]	--	--	[b]	--	--
Dieldrin	--	11	65	--	36	174
Aldrin	17	123	--	16	161	--
Chlordane	--	3	--	--	6	--
Endosulfan	--	61	--	--	56	--
Toxaphene	26,915	27,345	28,112	5,016	3,881	3,275
Others	2,660	85	--	428	285	--
Organophosphates	15,196	13,624	29,376	10,237	7,865	11,427
Disulfoton	565	300	225	619	473	553
Bidrin	--	1,857	778	--	1,416	1,797
Methyl Parathion	8,760	7,279	22,988	5,420	3,577	6,384

TABLE 3-7 (Continued)

Type of Insecticide Product	Pounds of Active Ingredients (x1,000)			Acres Treated (x1,000)[a]		
	1964	1966	1971	1964	1966	1971
Parathion	1,636	2,181	2,560	751	860	682
Malathion	1,811	559	670	213	245	273
Demeton	47	--	--	322	--	--
Trichlorfon	--	963	144	--	512	191
Azinophosmethyl	250	200	288	641	222	119
Phorate	10	--	100	35	--	182
Ethion	--	73	6	--	26	30
Other	2,117	212	1,617	2,236	534	1,216
Carbamates	4,524	1,571	1,291	1,002	415	394
Carbaryl	4,510	1,571	1,214	1,002	415	244
Methomyl	--	--	40	--	--	84
Other	14	--	37	--	--	66
Other organic	6	--	2	102	--	24
Total insecticides	78,022	64,900	73,357			

[a] The acres treated with a single insecticide may be overstated if different commercial pesticides containing the same ingredients are applied to the same acreage in separate treatments. Also, the number of acres treated with different ingredients should not be added together because two or more of these ingredients may have been used separately or together on the same acre.

[b] Less than 500 pounds or less than 10,000 acres treated.

SOURCE: USDA 1965a, 1967, 1972. *Quantities of Pesticides Used by Farmers in 1964*, Agricultural Economic Report #131, ERS, USDA; *Quantities of Pesticides Used by Farmers in 1966*, Agricultural Economic Report #179, ERS, USDA; *Quantities of Pesticides Used by Farmers in 1971*, (mimeo) #252.

Crop protection is one of many essential components of profitable crop production. The amounts of insecticide used on cotton and the cost involved in their purchase and application are high when compared with other field crops. For this reason many crop protection specialists and environmentalists have an exaggerated view concerning pest control and pesticide usage on the crop. A better perspective might be gained by comparing costs of insect control on the crop with the total costs of cotton production. The average percentage of total production costs for the insecticides and fungicides used on cotton from the period 1966-1969 ranged from 317 to 4 percent. Much larger percentages were spent for power and equipment, ginning, land, and general overhead than for crop protection (USDA 1972; Smith, 1971).

As shown in Table 2-2, the average per acre cost to cotton producers for insecticides in 1969 ranged from a high of $35.99 in southern California-southwest Arizona to a low of $0.53 in the Texas High Plains. The high cost areas are the boll-weevil-infested portions of the Cotton Belt and the Arizona-southern California deserts where the pink bollworm has become a serious problem since 1966. Boll weevil control costs are higher in high-rainfall areas of the Delta States and the Southeast than in the semiarid areas of Texas and Oklahoma (USDA 1972) where the pest is less favored by the climate.

The drastic effect on production costs of a major pest that must be controlled by insecticides is well illustrated by comparing insecticide costs in the pink-bollworm-infested areas of California-Arizona or the boll-weevil-infested areas of the Southeastern States with insecticide costs in the Texas High Plains. In most years, cotton farmers in the High Plains produce a crop without applying any insecticides.

The amount of insecticide used on cotton also is closely tied to the productivity of the land. Areas with high yield potential use much greater amounts of insecticide than areas with low yield potential. Thus, insecticide costs are much higher in the Delta States than in the dryland areas of Texas, although cotton in both areas may suffer heavy boll weevil infestations. Average per acre production in the Delta States is more than twice that of the Texas drylands, but insecticide costs per acre are more than five times greater (Reynolds et al. 1975).

Cotton producers apply insecticides to avert capital investment risks and to attempt to ensure profits. The greater the investment or potential profit, the more likely insecticide use becomes when there is a chance of loss induced by pests. Some producers treat their cotton with insecticides from planting to harvest as a preventive measure, regardless of pest insect numbers. Others treat only when pest numbers appear to be sufficient to damage the crop. Still others use a variety of pest control techniques including crop rotation, crop diversity, planting date changes, selection of early-maturing varieties, restricted use of irrigation and fertilizer, stalk destruction, and plowing, as well as insecticide application. Treatment on a preventive basis (from planting to harvest) generally is the most expensive control strategy. Using a variety of techniques is usually the most profitable strategy. Implementation of the most effective and profitable pest control methods depends on the managerial competence of the farm operator, available technical assistance from extension and private entomologists, and a sound economic base for the total farm operation. Often, insecticidal crop treatment on a particular farm can be reduced by increasing producer knowledge and confidence regarding new alternatives for insect pest suppression.

Impact of Insecticide-Resistant Pest Strains

All of the major arthropod pests of cotton have developed resistance to one or more pesticides (see Table 3-8).

The development of resistance to the boll weevil in the mid- to late-1950s provided the first impetus for cotton producers to switch from the chlorinated hydrocarbon to the organophosphorus and carbamate insecticides. Development of bollworm and tobacco budworm resistance to the chlorinated hydrocarbons caused an even greater shift to the OP compounds. This, coupled with restrictions on the use of DDT and similar materials in California and Arizona (because of drift hazards on other crops) had forced most cotton producers between Texas and California to rely on the OP compounds (often combined with toxaphene). These shifts occurred well before the EPA banned the use of DDT on cotton. The EPA action, combined with the earlier development of resistant pest strains, has nearly ended the use of chlorinated hydrocarbons on cotton. Toxaphene, a chlorinated camphene, and endrin are the only exceptions. Large amounts are still applied to cotton in formulations containing methyl parathion (Reynolds et al. 1975).

Regulatory actions other than restrictions or bans can have a great effect on the amounts of insecticide needed for cotton production. Texas, Arizona, and California have enacted legislation requiring farmers to plant, grow, and harvest cotton during a restricted season as a means for suppressing the pink bollworm. When these cultural practices are enforced in a large geographical area the need for insecticides for control of the pink bollworm (and often the boll weevil) may be greatly reduced (Adkisson and Gaines 1960).

TABLE 3-8 Arthropod Pests of Cotton That Have Developed Significant Resistance to Insecticides

Chlorinated Hydrocarbons	Organophosphorus Compounds	Carbamates	Arsenicals
Boll weevil	Bollworm	Bollworm	None
Bollworm	Tobacco budworm	Tobacco budworm	
Tobacco budworm	Spider mites (*Tetranychus* spp.)		
Pink bollworm			
Cotton fleahopper			
Lygus			
Cotton aphid			
Thrips (*Frankliniella* sp.)			
Cabbage looper			
Cotton leafworm			
Cotton leaf perforator			
Salt-marsh caterpillar			
Beet armyworm			

SOURCE: Adapted by permission from L.D. Newsom and J.R. Brazzel (1968) "Pests and their control," in Fred C. Elliot, Marvin Hoover, and Walter K. Porter, Jr. ed. *Advances in Production and Utilization of Quality Cotton*, Iowa State University Press, Ames, Iowa.

ENVIRONMENTAL EFFECTS OF CURRENT PEST CONTROL PRACTICES

Insecticides

Although many different insecticides are applied to cotton, toxaphene and methyl parathion represent 38 percent and 31 percent, respectively, of the total amount of insecticides used on cotton in 1971. These chemicals and DDT accounted for 88 percent of the insecticides sprayed on cotton in 1971. Since DDT has been banned, toxaphene and methyl parathion may contribute over 80 percent. Other chemicals (e.g., endrin) were important locally. In evaluating the effects of these chemicals, it is important to remember that the mix keeps changing because of pest resistance and the appearance of new chemicals.

The major environmental effects of the heavy insecticide use in cotton include fish kills, and mortality and reduced reproduction in other wildlife. These have been reported widely following aerial spraying of toxic organophosphates, but are also caused by some organochlorides. No systematic data estimating the overall magnitude of these effects are available.

The following account is based largely upon the recent EPA *Toxaphene Status Report*. Toxaphene is a chlorinated hydrocarbon insecticide that is in the same subgroup as endrin, but is closer to DDT in its environmental impact characteristics. It is moderately toxic to rats, being more toxic than DDT, but less toxic than aldrin, dieldrin, and endrin. Toxaphene is more acutely toxic to mammals, birds, fish, amphibia, and invertebrates than DDT. Birds that eat fish suffer unusually high mortalities after feeding in marshes containing toxaphene drained from surrounding farmland. Toxaphene occurred in the marsh fish at about eight parts per million (ppm). Another study revealed that water birds showed severe reproductive depression after a marsh had been treated with toxaphene at two pounds per acre (105 ppm in water). Toxaphene is highly toxic to fish and may persist in lakes for extended periods, sometimes preventing restocking of the waters for several years. It is also highly toxic to a wide range of invertebrates.

Toxaphene persists in the environment for long periods--its "half-life" in soil varies from 3 to 10 years and in water varies to 6 years (*Toxaphene Status Report*). Toxaphene is accumulated from water by aquatic plants, benthic invertebrates, and fish; however, bioconcentration (from a factor of 500 for aquatic plants to 20,000 for rainbow trout) is less marked than DDT. Some fish (mosquito fish, golden shiner, bluegill) can become toxaphene-resistant and, thereby, pass on increasing concentrations to larger fish-eating fish and to fish-eating birds. Insecticide contamination of runoff water from agricultural areas apparently is a major factor leading to the development of insecticide-resistant fish populations.

Toxaphene is highly toxic to a wide variety of predatory and parasitic insects. Whereas the major toxicity effects are upon pests in the cotton crop, presumably toxicity effects occur farther afield also, since toxaphene is persistent, subject to bioaccumulation, and is transported in water (including drainage water from fields) and perhaps in the atmosphere. However, the evidence to date suggests it does not appear as widely in the nation's waters as DDT and dieldrin.

Unfortunately, severe problems are involved in the chemical analysis of toxaphene and its derivatives. The EPA Toxaphene Report noted that "we have no significant information concerning the metabolism or degradation of toxaphene in the environment--neither physiological nor chemical data." Recently Casida et al. (1974) described studies defining the nature of toxaphene components and their metabolic fate in mammals.

Methyl parathion is among the most acutely toxic of insecticides. LD_{50} for various mammals is in the range 4 to 40 mg/kg and is lower for birds. There are many reports of domestic animal and wildlife mortality and poisonings following methyl parathion spraying of fields, but no quantitative data are available on the frequency or magnitude of these effects. Methyl parathion can be surprisingly persistent for an organophosphate (two years in water and five years in soil) and concentrates in food chains to a small degree.

Methyl parathion has directly impaired health of farm workers, owing particularly to its extreme dermal toxicity. Only in California (and only in recent years) has there been organized an attempt to measure the magnitude of this problem. As an example of the problem, in 1967, in the cotton growing counties of Cameron and Hidalgo in the Lower Rio Grande Valley in Texas (200,000 acres), 25 percent (57 out of 250) of the temporary workers who load spray planes reported acute toxicity effects. Other especially vulnerable groups are the applicators, field-workers, flagmen, and children and others exposed by accident. There is little doubt that pesticide poisonings are under-reported: loaders who stay away from work because they "do not feel well" are not counted as poisonings. One uncorroborated report from the San Joaquin Valley claims that in some cases farm workers suffering from pesticide poisoning have been entered in hospital records as "intoxicated."

In California, the Department of Public Health maintains records of illnesses related to pesticides that are more complete than in other states, though these records are still not fully adequate. It is not possible to tell from these statistics what fraction of cases is related to cotton production. Organophosphates, especially the parathions, are the most important contributors. The total reported incidence of occupational disease caused by pesticides in California during 1971 was 1,284. (These statistics were compiled by the State of California, Department of Public Health, based upon Doctor's First Report of Work Injury. See *Occupational Disease in California* 1972.) Comparison of the two hundred forty-seven cases of systemic poisonings in 1971 with the 583 cases for the first ten months of 1973 reveals an increase suggesting that farm worker safety is not improving in this respect.

Other Insecticides

Among the lesser used insecticides, parathion (2.5 million pounds on cotton in the United States in 1971, see Table 3-7) is even more hazardous than methyl parathion, while endrin (1 million pounds on cotton in the United States in 1971) is highly toxic to vertebrates, persistent, bioconcentrates to a high degree, and is easily dispersed throughout the environment. Endrin poses more serious environmental problems than toxaphene, on a unit weight basis, but is less heavily used in cotton.

The major insecticides used in cotton produce two major "external" types of deleterious impact. First, they cause an unmeasured loss of wildlife and disrupt ecosystems by direct killing of invertebrates and wildlife, suppressing reproduction, and through the development of resistance. Both toxaphene and methyl parathion cause these problems, though the greater persistence of toxaphene makes it the prime offender, except for local kills. Second, methyl parathion is a severe health hazard to farm workers, partly because it is difficult for the workers always to be sufficiently careful in handling the material. Parathion and endrin pose severe health and environmental problems but are relatively less heavily used in cotton.

Fungicides

The health and environmental effects of fungicides have not been examined to any extent, but the quantities used in cotton are small and appear not to produce serious hazards.

Herbicides

The use of herbicides in cotton has expanded rapidly over the past decade (Table 3-7). Study of environmental and chronic health effects of herbicides is less adequate than for insecticides, but more complete than for fungicides. A recent laboratory study (Lichtenstein et al. 1973) has shown that several herbicides can act synergistically to increase the toxicity of insecticides upon *Drosophila*, the housefly (*Musca domestica*), and larvae of the mosquito *Aedes aegypti*. Atrazine, simazine, monuron, and 2,4-D enhanced the toxicity of numerous organophosphates, two chlorinated hydrocarbons, and a carbamate. These herbicides are not used to any significant degree in cotton, but little work has been done on synergy. Such studies should be done.

Herbicides may be transported from cotton fields to nearby aquatic systems. Although no body of data exists to show widespread or severe environmental problems arising from the use of herbicides in cotton, in the absence of more study of the subject, we cannot conclude that they are environmentally harmless.

The environmental effects of other current pest control practices appear to be local and generally minor. Cultural activities such as plowing and cultivating can lead to local air pollution and to soil erosion which affects aquatic systems, including excessive siltation and eutrophication. Destruction of overwintering habitats for insects, weed destruction, and the trends towards larger fields and simpler farms increase the gradual reduction of wild flora and fauna variety, but these effects are confined mainly to the farm. The introduction or encouragement of insect natural enemies on cotton has no documented deleterious environmental effects that are known to us.

REFERENCES

Adkisson, P. L. (1966) Diapause boll weevil control: A comparison of two methods. Texas A & M Bull. 1054.

Adkisson, P. L. (1969) How insects damage crops. *In* How crops grow--A century later. Conn. Agric. Exp. Stn. Bull. 708:155-164.

Adkisson, P. L. (1972) The integrated control of the insect pests of cotton. Proc. Tall Timbers Conf. Ecol. Animal Control by Habitat Management. 4:175-188.

Adkisson, P. L. (1973) The principles, strategies and tactics of pest control in cotton. Mem. Ecol. Soc. Austr. 1:274-282.

Adkisson, P. L., and J. C. Gaines (1960) Pink bollworm control as related to the total cotton insect control program of central Texas. Tex. Agric. Exp. Stn. Misc. Publ. 444. 7 p.

Brazzel, J. R. (1961) Boll weevil resistance to insecticides in Texas in 1960. Tex. Agric. Exp. Stn. Prog. Rep. 2171. 4 p.

Brazzel, J. R. (1963) Resistance to DDT in *Heliothis virescens*. J. Econ. Entomol. 56:571-574.

Brazzel, J. R. (1964) DDT resistance in *Heliothis zea*. J. Econ. Entomol. 57:455-457.

Coad, B. R., and G. L. McNeil (1924) Dusting cotton from airplanes. U.S. Dep. Agric. Bull. 1204.

Crawford, J. L. (1974), Chairman, 1973 cotton disease loss estimate committee report, Proc. Beltwide Cotton Production Res. Conf., 34th Cotton Disease Council.

Cross, W. H., M. J. Lukefahr, P. A. Fryxell, and H. R. Burke (1975) Host plants of the boll weevil. Environ. Entomol. 4:19-26.

Eichers, T., P. Andrilenas, I. I. Blake, R. Jenkins, and A. Fox (1970) Quantities of pesticides used by farmers in 1966. U.S. Dep. Agric. Econ. Res. Serv. Agr. Econ. Rep. 179. 61 pp.

Falcon, L. A., R. van den Bosch, C. A. Ferris, L. K. Stromberg, L. K. Etzel, R. E. Stinner, and T. F. Leigh (1968) A comparison of season-long cotton pest-control programs in California during 1966. J. Econ. Entomol. 61:633-642.

Harris, F. A., J. B. Graves, S. J. Nemec, S. B. Vinson, and D. A. Wolfenbarger (1972) Insecticide resistance. In distribution, abundance and control of *Heliothis* species in cotton and other host plants. South. Coop. Ser. Bull. No. 169.

Lichtenstein, E. P., T. T. Liang, and B. N. Anderegg (1973) Synergism of insecticides by herbicides. Science 181:847-849.

Malley, F. W. (1901) The Mexican cotton boll weevil. U.S. Dep. Agric. Farmer's Bull. 130.

Newsom, L. D. (1972) Theory of population management for *Heliothis* spp. in cotton. Distribution, abundance, and control of *Heliothis* species in cotton and other host plants. South. Coop. Ser. Bull. No. 169:80-92.

Newsom, L. D., and J. R. Brazzel (1968) Pests and their control. Pages 367-405 *in* Advances in production and utilization of quality cotton: principles and practices, F. C. Elliot, M. Hoover, and W. K. Porter, Jr., eds. Iowa State University Press, Ames.

Occupational disease in California attributable to pesticides and other agricultural chemicals 1972. State of California, Department of Public Health.

Pimentel, D. (1973) Extent of pesticide use, food supply, and pollution. J.N.Y. Entomol. Soc. LXXXI:13-33.

Post, G. B. (1924) Boll weevil control by airplane. Ga. State College Agric. Bull. 301. Athens, Ga. 22 p.

Reynolds, H. T., P. L. Adkisson, and R. F. Smith (1975) Cotton insect pest management, pp. 379-443. *In* R. L. Metcalf and W. H. Luckman, eds. Introduction to insect pest management. John Wiley and Sons, New York.

Roussel, J. S., and D. F. Clower (1955) Resistance to the chlorinated hydrocarbon insecticides in the boll weevil (*Anthonomus grandis* Boh.). La. Agric. Exp. Stn. Circ. 41.

Smith, R. F. (1971) Economic aspects of pest control. Proc. Tall Timbers Conf. Ecol. Animal Control by Habitat Management. 3:58-83.

Smith, R. F., and H. T. Reynolds (1972) Effects of manipulation of cotton agroecosystem on insect pest populations. Pages 373-406 *in* M. T. Farvar and J. P. Milton, eds. The careless technology--ecology and international development. Natural History Press, Garden City, New York.

Smith, R. F., and L. A. Falcon (1973) Insect control for cotton in California. Cotton Grow. Rev. 50:15-27.

Sterling, W. L., and R. L. Haney (1973) Cotton yields climb, costs drop through pest management systems. Tex. Agric. Exp. Stn. Tex. Agric. Prog. 19:4-7.

Stern, V. M. (1969) Interplanting alfalfa in cotton to control lygus bugs and other insect pests. Proc. Tall Timbers Conf. Ecol. Animal Control by Habitat Management. 1:55-69.

Stern, V. M., R. van den Bosch, and T. F. Leigh (1964) Strip cutting alfalfa for lygus bug control. Calif. Agric. 18:4-6.

Thomas, F. L., W. L. Owen, J. C. Gaines, and F. Sherman, III (1929) Boll weevil control by airplane dusting. Tex. Agric. Exp. Stn. Bull. 394.

Townsend, C. H. T. (1895) Report on the Mexican cotton boll weevil in Texas (*Anthonomus grandis* Boh.). Insect Life 7:295-309.

U.S. Department of Agriculture (1965) Losses in agriculture. Agric. Handbook No. 291, U.S. Government Printing Office.

U.S. Department of Agriculture (1965a) Quantities of pesticides used by farmers in 1964. Agricultural Economic Report #131, Econ. Res. Serv.

U.S. Department of Agriculture (1967) Quantities of pesticides used by farmers in 1966. Agricultural Economic Report #179. Econ. Res. Serv.

U.S. Department of Agriculture (1972) Quantities of pesticides used by farmers in 1971. Agricultural Economic Report #252. Econ. Res. Serv.

U.S. Department of Agriculture (1972a) Extent and cost of weed control with herbicides and an evaluation of important weeds. ARS-H-1.

van den Bosch, R., T. F. Leigh, L. A. Falcon, V. M. Stern, D. Gonzales, and K. S. Hagen (1971) The developing program of integrated control of cotton pests in California. Pages 377-394 *in* C. Huffaker, ed. Biological control. Plenum Press, New York.

4

CURRENT TRENDS AND FUTURE ALTERNATIVES IN PEST MANAGEMENT

In most cases, chemical methods of pest control have been directed towards total destruction of cotton pests. Not surprisingly, the pests have adapted in various ways: (1) insect species have developed resistance to certain chemicals; and (2) herbicide use has sometimes resulted in the replacement of one major weed species by another. Thus, sole reliance on chemical control methods has not had the lasting success first envisioned. Scientific pest management requires a knowledge of ecological principles, the biological intricacies of each pest, and the natural factors tending to regulate their numbers. Cotton insect pests are well adapted to thrive in the crop and to adjust to changing conditions of crop production. Clearly, this means research objectives and pest control practices must be based on unified and balanced systems of widely proved principles of insect, weed, and disease control.

ARTHROPOD PESTS

Genetic Manipulation and Population Eradication

The introduction of dominant lethal mutations to a population can lead to a population collapse. This approach to pest population suppression or extinction has been advocated and applied in a few cases. The most notable example is the elimination of the screwworm, *Cochliomyia hominovorax*, from the southeastern United States following the massive release of irradiated males. In the Southwest, a similar effort reduced the screwworm population below levels of economic damage, but recent resurgences have occurred (Whitten and Foster 1975). In addition to chromosome damage induced by radiation, other genetic techniques that could be used to produce lethal factors, including inseparable association of meiotic drive with genes for female sterility, conditional lethal mutations, and unbalanced genetic equilibrium produced by translocations and compound chromosomes (Smith and von Borstel 1972). Using the characteristic of diapause determination, the potential of dominant conditional lethals for inducing the collapse of pest populations has been formulated by Klassen et al. (1970). In their lucid analysis, Smith and von Borstel (1972) pointed out that chromosomal translocations could be utilized as contrived dominant lethals, if the translocations were developed through a carefully designed breeding program.

> Two separate lines could be established, each homozygous for a number of chromosomal translocations. Matings between the two lines would produce offspring heterozygous for all the translocations. If each line contained three different translocations, over 98 percent of the gametes would be lethal. Such a genetically contrived dominant lethal method would be a substantial improvement over the radiation-induced dominant lethal method, since the lethal gametes would be produced over the entire life span of each animal.... This method could be used profitably in a species such as the boll weevil, *Anthonomus grandis*, which is strikingly sensitive to radiation and therefore can be killed at exposures less than that needed to obtain a high frequency of dominant lethality in the sperm.

Although highly promising as a technique to suppress pests, genetic manipulation requires that the target population be relatively low in number and restricted in area. This requirement comes from the fact that the lethal factor must be introduced in numbers sufficient to overwhelm the genetic composition of the target gene pool. As a result, genetic manipulation must be effected subsequent to other intensive programs designed to reduce drastically the pest population. The introduction of contrived dominant lethal factors is usually visualized as the last operation in a so-called eradication program.

Eradication programs including genetic manipulation are currently being directed against both the boll weevil and the pink bollworm, *Pectinophora gossypiella* (Knipling 1971, Cross 1973). In opening his review of the biology, control, and eradication of the boll weevil, Cross (1973) stated, "The boll weevil, *Anthonomus grandis*, is currently in the limelight as a superpest of cotton which may possibly be eradicated from the United States or even from its complete range of distribution."

Boll weevil eradication programs that have been tested on a pilot scale and proposed on a Cotton-Belt scale include successive application of population suppression measures: (1) intensive chemical control, during the growing season and especially "diapause control" late in the growing season as the weevils are entering diapause and seeking overwintering sites (Adkisson 1966); (2) pheromone trapping in the spring to further reduce the population; (3) insecticide-treated trap crop plantings, also with pheromone traps; (4) a single insecticide treatment of cotton as the plants begin to fruit; and (5) massive introduction of sterile males or other dominant-lethal-carrying individuals. Technical shortcomings, particularly in the degree of sterility and behaviorial competitiveness of the released "sterile" weevils seriously limited the effectiveness of the pilot eradication experiment (Whitten and Foster 1975, Eden et al. 1973). The boll weevil population, however, was largely suppressed, demonstrating the effectiveness of the methodology in pest management.

The absolute eradication of a major pest species such as the boll weevil would be of economic benefit in terms of both dollar outlay and environmental protection from continued pesticide usage. Present technology is equal to the task of eradicating insular populations and populations that are severely limited by geographical barriers. Eradication of nonisolated, widely distributed, continental populations appears beyond the capability of current technology.

The wisdom of a program to eradicate the boll weevil from the United States alone has been questioned, partly on economic grounds and partly because of biological considerations. The population models on which eradication programs are based (Knipling 1966, 1971, 1972a, 1972b) involve the assumption that the genetic constitution of the target population will remain relatively constant during the eradication process. That is, it is assumed that there will be no natural selection countering the pressures imposed on the population. This premise may not be justified. There may be qualifying factors and selection pressures leading to changes in gene frequencies roughly similar to those that led to the appearance of resistance to insecticides. It has been suggested that such changes are likely (Smith and von Borstel 1972), and may have been involved in recent screwworm population resurgence (Whitten and Foster 1975). The existence of alternate wild host plants, e.g., *Cienfuegosia* spp. (Cross et al. 1975), will further complicate complete or partial eradication of the weevil over such a massive continental area.

The value of the population suppressive techniques discussed above should be very great in the management of the boll weevil and of other insect pests affecting cotton. However, the proposal that these techniques (some of which are not yet fully developed) be pursued toward the goal of absolute eradication of any major continental species does not appear to be justified at present, on either economic, technical, or biological grounds.

Cultural Practices

Present technology is sufficient to greatly reduce the amounts of insecticides needed for protection of cotton in the United States. This can be accomplished without any reduction in yield or quality. In areas where pest strains resistant to insecticide are rampant, yields may even be increased by using nonchemical methods to suppress insect pest numbers in an integrated program with carefully supervised insecticidal chemical applications used only as needed to control outbreaks. The main principles involved in the alternate methods involve manipulation of the environment: (a) to make it as unfavorable as possible to the pest species (or more favorable to entomophagous species); (b) to reduce the rate of pest increase and damage; or (c) to concentrate pest numbers in small areas where direct control measures may be applied with minimum disruption of the entomophagous species (Isely 1948, NAS 1969, Stern et al. 1975, Adkisson 1974).

Cultural Practices Reducing Overwintering Pest Numbers

Short growing periods and early harvest and disposal of crop residue have long been recognized as excellent measures for reducing numbers of potential overwintering pest insects. The practice removes food and breeding sites before environmental conditions force the insects into diapause. Short growing periods allow fewer pest generations to develop and may cause the crop to be unsuitable as a food source for the pests at a crucial time in their seasonal history. In addition, it makes possible the early destruction of food sources and breeding sites by direct measures, such as chemical defoliation of plants, stalk shredding, and plow under of crop remnants. The first entomologists in the United States to conduct research on the boll weevil and pink bollworm recognized the value of these measures for controlling the two pests and recommended their implementation on an areawide basis (Malley 1902, Hunter 1912, Ohlendorf 1926, Noble 1969).

The present cultural program for the control of the pink bollworm evolved from a series of studies which provided a thorough understanding of the seasonal history of the species. The eggs of the pink bollworm are laid in protective sites on the fruiting forms of cotton. The larvae, immediately after hatching, burrow into the squares or bolls. Because of this behavior, they are difficult to control with insecticides and relatively safe from parasites and predators (Reynolds et al. 1975).

Early research by Ohlendorf (1926) showed that the pink bollworm diapauses as a last instar larvae and generally overwinters in the seed of bolls left in the field after harvest. This is the "weak-link" which is vulnerable to attack. Later studies showed that the diapause is controlled by the photoperiod (Lukefahr et al. 1964, Adkisson et al. 1963). The first diapause of the year always occurs during early September when day lengths are less than 13 hours (Adkisson 1964). The incidence of diapause then increases rapidly and attains a maximum in mid-October and early November. This response to photoperiod provides the key to control because the onset of diapause may be predicted for any given location with great precision (Adkisson 1966).

The timing of cultural practices in cotton production was modified to take advantage of this phenomenon so as to maximally reduce the size of the overwintering pink bollworm population. The sequence of cultural practices from the time of crop maturity in the fall to the planting of the subsequent crop in the spring is as follows: (1) defoliate or desiccate the mature crop to cause all bolls to open at nearly the same time, expediting machine harvesting; (2) harvest the crop early, shred stalks, and plow under crop remnants immediately; (3) irrigate before planting in desert areas if water is available; and (4) plant new crops during a designated planting period that allows for maximum suicidal emergence of overwintering moths, i.e., moths that emerge and die before cotton fruit is available for oviposition (Adkisson and Gaines 1960).

Early-maturing varieties are important to the success of this program. Development of most of the potential overwintering population of pink bollworms can be prevented if the cotton is defoliated or desiccated in late August or early September, rather than in October or November (Adkisson 1962). This expedites the early and rapid harvest of the crop by mechanical harvesters. Mechanical strippers leave fewer larvae in the field than spindle-pickers. Almost all larvae carried to the gin in seed-cotton are killed by the ginning process (Robertson et al. 1959). A spindle-picker may leave immature bolls in the field to harbor sources of infestation for the following year. However, if the stalks are shredded and plowed under, more than 90 percent of the larvae may be killed (Wilkes et al. 1959, Adkisson et al. 1960, Noble 1969, Fenton and Owen 1953).*

The development of diapausing boll weevils can be terminated or prevented by many of the measures used against the pink bollworm. Malley (1902) had developed sufficient information to promote areawide destruction of cotton in early fall as the best means of weevil control before effective insecticides had been discovered for control of the pest. Hunter (1912) also reported that farmers in a Calhoun County, Texas, experiment reduced the weevil to such low numbers by uprooting and burning cotton stalks in 1906, that a good crop was produced in 1907 without use of any other control measures. Annand (1948) later reported results of a three-county demonstration in the Lower Rio Grande Valley of Texas in 1945 when practically 100 percent of the stalks were shredded and plowed under before September 1. The following year, boll weevil numbers did not reach economically significant proportions in the Valley and no insecticides were used for their control even though boll weevils inflicted severe damage to cotton in other parts of Texas.

Since these early demonstrations, considerably more has been learned about the diapause of the boll weevil. Unlike the pink bollworm, the boll weevil diapauses as an adult and hibernates outside the cotton field. Thus, to obtain maximal reduction of potential overwintering numbers, the prehibernating weevils must be killed before they leave the cotton fields for hibernation in nearby

---

*This combination of practices has reduced the pink bollworm in Texas to the status of a minor pest and insecticides are seldom needed for control. The value of the program is such that regulatory measures have been enacted by the Texas Legislature to force compliance by cotton producers on an areawide basis.

woody or bushy sites. This can best be accomplished by modifying the pink bollworm control program to include selective use of an OP insecticide during the harvest period. The insecticide is added to the defoliant or desiccant just prior to harvest. One or two additional insecticidal treatments may be required if the harvest season is prolonged (Reynolds et al. 1975).

This program has proven highly effective in reducing boll weevil numbers without adverse impact on the arthropod natural enemies of the bollworm or tobacco budworm (Brazzel 1961, Brazzel et al. 1961, Adkisson et al. 1966). In many years this combination of practices, applied areawide, has so reduced boll weevil numbers that they were not able to attain economically significant proportions during the subsequent growing season. Also, because it conserves insect natural enemies, this program has been successful in averting economically important outbreaks of the bollworm and tobacco budworm. Implementation of these measures has greatly reduced the amounts of insecticides needed to produce cotton with a concomitant increase in profit to the producer (Adkisson 1972, Sterling and Haney 1973).

Pheromone Manipulation and Trap Crops

Practices discussed in the previous section are capable of greatly reducing overwintering boll weevil numbers, but they may not always suppress the population to a level needed to prevent economic damage to cotton. One technique that shows considerable promise is the use of the boll weevil pheromone, "grandlure," (Keller et al. 1964, Hardee et al. 1969) to bait limited areas of cotton fields thereby creating spring trap crops for overwintering adults. These weevils, then, may be killed in the trap crop with a limited amount of insecticide before they reproduce. This practice does not upset the entomophagous species inhabiting the remainder of the field. Bottrell (1972) in Texas has been able to maneuver more than 70 percent of the overwintering boll weevils that entered a cotton field into a trap crop occupying less than 15 percent of the field area. The weevils then may be destroyed by insecticidal treatment, shredding, and plowing under of the trap. This can be accomplished with little impact on the natural enemies of the bollworm and tobacco budworm. Lloyd et al. (1972) reported similar results in Mississippi. In addition, Bottrell has had limited success using the pheromone to concentrate hibernating weevils in restricted areas of their overwintering habitats. There they may be destroyed by mechanical or physical means.

These tactics have not been developed to the point where they may be recommended to farmers. However, initial success in large research plots has been excellent. These methods show great promise for practical use in further reducing overwintering boll weevil populations while conserving the natural enemies of other cotton pests.

Short-Season and Resistant Varieties

For many years, varietal modification has been believed to offer one of the most effective means for reducing damage that insects inflict on cotton. Malley (1902) sought early-maturing varieties while Isely (1928, 1934) sought both early-maturing and insect-resistant varieties. However, it is only in recent years that entomologists and plant breeders have been able to develop breeding stocks of cotton that offer real promise for pest insect suppression. Maxwell et al. (1972) have made an intensive review of the literature in this field; therefore, only a few pertinent examples will be noted in the present discussion.

The typical commercial cotton variety is indeterminate in fruiting. It has little, if any, resistance to pest insects and may require insecticidal protection for long periods during the season. New cotton types being developed by J. A. Niles, L. S. Bird, and J. K. Walker of the Texas Station in cooperation with M. J. Lukefahr and Robert Dilday of the USDA carry characters conferring some resistance to the cotton fleahopper, bollworm, and tobacco budworm (Lukefahr et al. 1970, Walker et al. 1972). In some production areas semi-dwarf lines adapted to high-density planting fruit rapidly over a short period and mature three to four weeks earlier than the present commercial types.

One selection nearing release in the Texas-Oklahoma area fruits so rapidly over such a short period that only about 50 percent of the fruit set is vulnerable to attack from the boll weevil (Walker and Niles 1971). This selection literally "out-runs" the buildup of the boll weevil. In addition, it matures so early that it should provide "automatic" control of diapausing boll weevils and pink bollworms simply because the crop may be harvested, and the stalks shredded and plowed under before environmental conditions are appropriate to force the pests into diapause.

Since the cotton can be grown in a system in which insecticidal control of the boll weevil may not be needed during the early mid-growing season, bollworm and tobacco budworm attacks should be minimized.

A frego-bract cotton variety has been developed that provides an extremely high level of resistance to the boll weevil (Jenkins and Parrot 1971). Unfortunately, the variety is susceptible to attack by the cotton fleahopper and *Lygus* spp. However, Schuster and Maxwell (1974) have developed a nectarless variety that is moderately resistant to lygus and fleahoppers and slightly resistant to *Heliothis* spp. This variety should be in commercial production in 1974 or 1975. It seems possible that the nectarless and frego-bract characters may be combined to produce a stock resistant to the boll weevil, fleahopper, lygus, and heliothis.

Resistant and short-season cotton varieties will have their greatest utility in integrated control programs utilizing cultural, chemical, and biological pest suppression techniques. In the boll-weevil-infested areas of the Cotton Belt, these varieties offer the necessary ingredient for developing pest management schemes that could greatly reduce, and perhaps even eliminate, present dependence on insecticides for control of the principal arthropod pests of the crop.

Other Cultural Practices

In the San Joaquin Valley of California, lygus bugs (principally *Lygus hesperus* Knight) are the major pests of cotton. Insecticidal treatments for the control of lygus may unleash secondary pest attacks by the bollworm, spider mites, *Tetranychus* spp., beet armyworm, *Spodeoptera exigua* (Hubner), cabbage loopers, *Trichoplusia ni* (Hubner), and other defoliators. Control of these secondary pests by insecticides is difficult and costly. In addition, the damage inflicted by secondary pests may exceed the loss that would have been caused by uncontrolled lygus (Falcon et al. 1968, Smith and Falcon 1973, Stern 1969). The lygus build their numbers in two crops, alfalfa and safflower, where they are not considered pests. When the alfalfa is cut, or the safflower matures, the lygus move to cotton. There they may cause significant yield losses if not controlled. One alternative to insecticidal control of lygus in cotton is to stabilize the environment in the alfalfa in order to keep the lygus in a crop which they do not damage (Stern 1969). Another practice may be to selectively treat the safflower with one insecticidal treatment at the precise time required to prevent movement of lygus to cotton (Mueller 1971).

In a large part of the San Joaquin Valley, alfalfa is grown in fields adjoining cotton. If an entire alfalfa field is mowed at one time, the lygus move into the cotton in great numbers. However, if the alfalfa is strip-cut so that two different age alfalfa growths are maintained simultaneously in the same field, the dispersal of the lygus to cotton may be prevented. When one set of alfalfa strips is cut, the half grown alternate strips are favorable to the lygus. By this technique, the lygus population can be manipulated from one alfalfa strip to the other and the numbers in cotton held below the economically significant threshold (Stern et al. 1967).

In areas where cotton is grown in huge blocks, it is recommended that 20-foot wide strips of alfalfa be interplanted at intervals of approximately 500 feet across the cotton field. The alfalfa is strip-cut to maintain a favorable environment for the lygus in part of the strip. This system, when properly managed, has virtually eliminated the need for insecticidal treatment of lygus in the interplanted fields. By conserving insect parasites and predators, the elimination of insecticidal treatments for lygus has averted outbreaks of secondary pests (Stern 1969).

Biological Control

The importance of entomophagous insects in cotton was recognized by early entomologists working with the crop. Gorham (1847) was aware of parasitism of the cotton leafworm; Riley (1873) listed many species of parasites and predators still important today. Before the boll weevil invasion of the United States, most of the insects that attacked cotton occurred in such low numbers that they never became pests of any consequence. Occasionally, natural enemies would fail to control one of these species and an outbreak would occur. An exception was the cotton leafworm. This annual migrant from Central America frequently inflicted serious damage to cotton. Control was achieved with arsenical insecticides (Reynolds et al. 1975).

When the boll weevil invaded the United States, an intensive search for parasites and predators of this pest was begun. More than 50 species of parasites and predators were discovered (Pierce 1908, Pierce et al. 1912). However, it soon became apparent that natural enemies alone were not capable of controlling the boll weevil.

Cotton growers turned to cultural methods and insecticides for crop protection.

Entomologists have not been able to quantify with precision the effectiveness of arthropod parasites and predators in keeping cotton pest species below economic threshold densities. With the widespread use of calcium arsenate, it became apparent immediately that there were many effective insect natural enemies operating in cotton. Arsenical treatments of cotton were followed closely by outbreaks of aphids and bollworms (Folsom 1928, Dunnam and Clark 1941, Ewing and Ivy 1943). These outbreaks were found to result from the loss of predation in treated fields.

Still, the value of natural biological control was not fully appreciated by entomologists, or cotton producers, until after the introduction of the synthetic organic insecticides. Widespread application of these chemicals elevated many insect species, viz, the bollworm, tobacco budworm, cotton leaf perforator, spider mites, cabbage looper, beet armyworm, and salt marsh caterpillar, to the status of major pests. In fact, the bollworm and tobacco budworm now probably inflict more damage to the crop each year than the boll weevil (Whitcomb 1970, Newsom 1970, Hagen et al. 1971, van den Bosch et al. 1971, Adkisson 1969).

More than 600 species of predators and parasites have been found in cotton, where they do an effective job of keeping more insect pests below crop damaging numbers (Whitcomb and Bell 1964). However, they cannot be depended on to control all cotton pests at all times. (They cannot be expected to control the boll weevil at any time.) Often only a single application of insecticide on cotton reduces the effectiveness of the natural enemies to the point that a secondary pest outbreak is allowed to occur (Newsom and Brazzel 1968). Thus, crop protection specialists presently are searching for control methods that conserve natural enemies. Most approaches minimize the use of insecticides or time applications to cause minimum disruption to the natural enemy complex.

Two other methods of utilizing insect natural enemies are importation of exotic species and augmentation of native species. Importations of exotic species have not been particularly successful, although some entomologists (Newsom and Brazzel 1968) believe there is great need to continue the search in Central America and Mexico for parasites of the boll weevil. Augmentation of naturally occurring parasites and predators by programmed releases of insects reared in laboratories has provided satisfactory control of certain pest species, particularly *Heliothis* spp., in experimental plots (Ridgway 1969, Lindgren 1969). However, techniques for large-scale rearing and release of insect natural enemies are not yet developed to the point that augmentation of natural populations is practical for grower use.

Epizootics of naturally occurring insect diseases are frequently observed. In fact, in many areas of the Cotton Belt, producers commonly depend on the viral diseases of the cabbage looper for control of the pest. The cabbage looper virus is more efficient than most insecticides and provides control of the pest when numbers reach a sufficient density. Generally, loopers are wiped out before damage which reduces yield occurs.

Insects also are attacked by a number of other microorganisms including bacteria, protozoa, fungi, rickettsiae, and nematodes. All these pathogens play important roles in the natural regulation of insects and mites. Besides causing death, pathogens may interfere with insect development, alter reproduction, lower resistance to attack by other microorganisms, or increase susceptibility to insecticides (Falcon 1971).

Among pathogens known to be important to the insect pests of cotton, *Baccillus thuringiensis* (BT) and the nuclear polyhedrosis viruses (NPV) of the bollworm, cabbage looper, and beet armyworm are of the most interest to researchers and commercial firms. *B. thuringiensis* is registered for use on cotton and is available commercially. The NPV of *Heliothis* spp. has been labelled experimentally and widely tested on cotton. However, its continued use is dependent on protocols for labelling that are to be established by the EPA.

Commercial preparations of BT and NPV can be applied in dusts or sprays in the same manner as chemical insecticides. Often, the control attained has been as good as that produced with insecticidal treatments, but it is not as consistent (Ignoffo 1970, Newsom and Brazzel 1968). Before insect pathogens are to be used on a wide scale for pest control in cotton they must: (1) produce more reliable and consistent control of pest species; (2) be more economical to use; (3) be placed in formulations that have better shelf-life, are easier to apply, and are more persistent in the field than present formulations; and (4) meet registration standards established by the EPA.

The potential importance of pathogens for control of the insect pest of cotton appears to be great. Their greatest utility will be in integrated control programs involving the use of other suppressive measures. However, much research yet must be done on efficiency, safety, production, formulation, and application before these microbial pesticides will be of much practical use to cotton growers (Falcon 1971).

In evaluating alternative control techniques for the future, it must be realized that, in principle at least, insect pests can evolve in response to almost any technique. For example, insect pests may shorten development time in response to short season varieties or may overcome varietal resistance. We are not able to predict which techniques are likely to suffer most in the future from this capacity of insect pests, though biological agents do have the potential advantage that they can evolve as the pest evolves, thus, "keeping up" with the pest.

WEEDS

The use of herbicides for weed control does not eliminate the need for sound cultural control practices and other preventive measures. In fact, the most reliable weed control programs use many methods.

Nonchemical alternatives have been developed and are in use for some weed situations. However, given current cotton production practices, alternative herbicides are needed as much or more than alternative nonchemical weed control methods. Alternative chemicals are needed to fit the shifting weed conditions throughout the Cotton Belt.

Alternatives to Current Practices

Biological control of weeds is occasionally mentioned as an alternative to herbicide use in cotton. If suitable and effective biological agents were available, this method would be desirable. The objective of weed control is to maintain specific weed populations below the level at which they are damaging to the crop. However, this desired population level may be below the level at which the biological control agent and the weed species are in balance. Furthermore, the classification of a given plant as a weed varies. A biological control agent may move from an area where the plant is considered a weed to areas where the plant is considered valuable. For example, Johnson grass is a recommended forage crop in some areas of the South and a noxious weed pest in other areas. Because of the risk to crop and ornamental plants, unspecialized insect feeders could not be used. Many weeds are closely related to specific crops (Johnson grass and sorghum, spurred anoda and cotton, horse nettle and tomato). Biological control is a specific and selective form of weed control and might be effective against a single weed species. However, if one specific weed species is eliminated from an area, other weed species are likely to become dominant. Thus, any controlling factor which concentrates on a single species is not likely to be a long-term solution to weed problems in cotton. In addition, the use of insecticides on cotton would kill any arthropod biological control organism. This would not affect pathogenic control organisms, unless the pathogen is transmitted by an insect. Even so, if future insect control programs on cotton do not use insecticides, then biological control of weeds might become more feasible.

Biological control of weeds in crops has been given limited study. For example, perhaps biological agents that either attack weeds or compete with weeds, but not with cotton plants, could be developed. Funding of research on such biocontrols has been too limited to date for proper evaluation of its future potential.

There are other potential weed control measures under development, including the use of mechanical procedures such as microwave or ultrasound. Current evaluations indicate that microwave machinery is very heavy and the system is slow. Only shallow control of weeds is achieved. Weeds germinating in the soil deeper than one inch have been able to survive and grow successfully. Thus, the system might cause an ecological shift away from shallow germinating weed species toward deeper germinating weeds that are more difficult to control by both machines and herbicides. There is also the disadvantage that once the microwave machine goes over a field, the soils cannot be further disturbed in preparation for planting. Any tillage would disturb the soil and bring new weed seeds to the surface. Chemicals which neutralize weed enzyme systems are also being studied as a potential means of increasing weed susceptibility to commonly used herbicides. However, work in this area has shown little promise to date, primarily because weed enzyme systems are so similar to crop enzyme systems.

Integrated Weed Control Practices

Most weed scientists advocate multiple approaches to weed problems. It is recognized that single measures undertaken to control a given weed pest often fail. However, weed control problems can often be ameliorated by applying a series of coordinated techniques which have a greater impact than any single component. The complementary action of various components in an integrated weed control program is illustrated by a study which showed flame weeding reduced hoe labor in cotton by 11 hours per acre, and use of a preemergence herbicide reduced hoe labor by 18 hours per acre. However, when both procedures were used, the hoe labor requirement was reduced by 47 hours per acre. In fact, a series of weed control procedures, including crop and herbicide rotations, different types of land preparation and cultivation, and other improved agronomic practices, are most effective as weed control measures when integrated with a complete farm management system.

The weed science phase of pest management emphasizes the concept of total farm weed control. If weeds are allowed to grow in other crops, set aside acres, or fence rows, weed seed populations in the soil cannot be decreased. A reduction in weed seed populations can reduce costly annual applications of other weed control measures. Increased emphasis should be placed on the development of integrated weed control programs for the entire farm.

PLANT PATHOGENS

Alternatives to chemical methods of controlling seedling diseases could include: (1) breeding or selection for resistance; (2) planting when soil temperatures are optimal for rapid emergence; and (3) development of biological control methods. Breeding or selection for resistance has been successful in many cases. However, resistance has not been developed to *Pythium* or *Rhizoctonia*. Late planting, while providing disease control, often conflicts with other desirable production and pest control practices. Biological control methods might involve the use of antagonistic fungi in the soil. When added to sterile soil, some antagonists have been shown to control *Rhizoctonia*. But when added to natural soil, they were gradually reduced to ineffectiveness by the equilibrium effects of natural soil microflora. Seed treatment with spore-forming bacteria, such as *Bacillus subtilis*, may have potential as a control practice for the future.

ENVIRONMENTAL EFFECTS OF ALTERNATIVES TO PRESENT CONTROL PRACTICES: INTEGRATED PEST MANAGEMENT CONTROL PROGRAMS

Weed control in the future is likely to be characterized by heavier use of herbicides and a greater variety of mixes. Although there is disagreement on this point, weed resistance may occur in the future.

The major changes in techniques and in the mix of techniques are likely to occur in insect pest control, with trends toward reduced use of chemicals and increased use of cultural and biological control methods. Cultural techniques, relying upon shorter growing seasons, synchronized harvest, crop residue destruction, and so on, may slightly increase the use of defoliant chemicals and increase local soil erosion problems, but these are likely to have only local effects. Pheromones and hormones might have minor effects on some populations of insects close to cotton fields, but no major problem is foreseen. Similarly, biological control should present no environmental problems except that other insect populations may be adversely affected, and the use of pathogens has not yet been sufficiently studied to allow their general use.

MODELLING AND SYSTEMS ANALYSIS

Modelling can serve as a mechanism for bringing together various aspects of a problem and for producing a predictive output. Systems analysis techniques can be effectively applied to the development of integrated pest control practices, especially the establishing of economic thresholds. An economic threshold is defined as the minimum pest density at which a pesticide treatment is justified by a comparison of treatment costs and potential losses (Stern 1966). The value of the crop losses that would be incurred if the pesticide were not applied should be greater than the cost of the treatment. The economic threshold concept is an important advance in pest control research because it recommends tolerance of pest populations that are of significant density but are below the economic threshold (Figure 4-1). The density at which treatment is economically justified depends upon pest

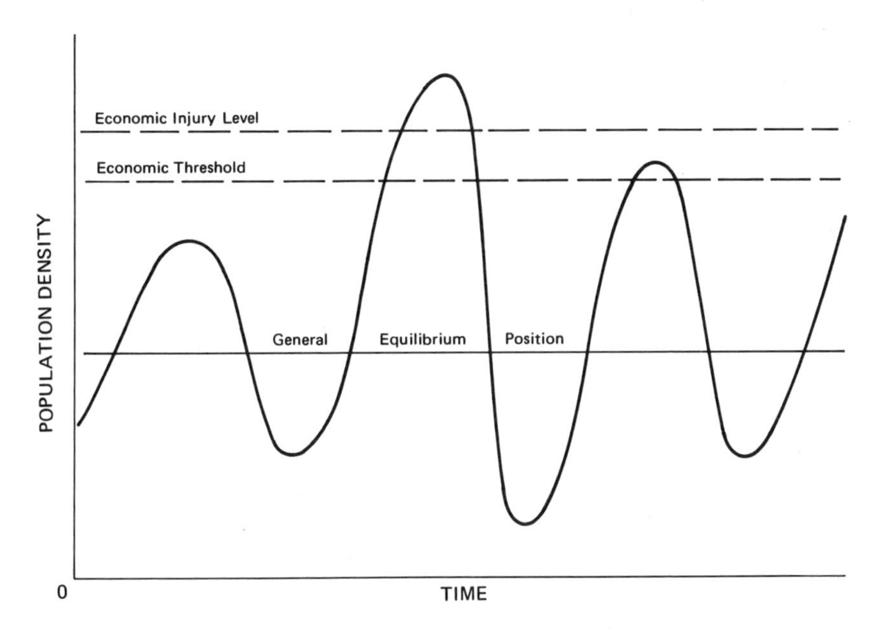

FIGURE 4-1. Schematic graph of a theoretical arthropod population over a period of time and its general equilibrium position, economic injury level, and economic threshold. (Source: Stern 1966.)

population size, its age structure, natural enemy population, time, crop maturity, and stress from lack of water or nutrients. In an attempt to define multi-dimensional economic thresholds, i.e., thresholds which depend on many variables, agriculturalists have begun to use mathematical models of the agroecosystem.

The general approach to the systems analysis aspect of integrated pest management has been to build simulation models for each of the major components (pests, crop plants, and natural enemies) of the agroecosystem being studied. Then the separate models can be simulated simultaneously by a computer to estimate the reaction of the ecosystem to various exogenous factors such as weather and pesticide applications.

Insect simulation models usually divide the population into a number of age compartments. The flow rates between compartments depend upon temperature, age, mortality, and fecundity factors. An example of this approach is the model of *Heliothis zea*, the cotton bollworm and corn earworm, developed by Stinner et al. (1974). Figure 4-2 gives a schematic description of this model.

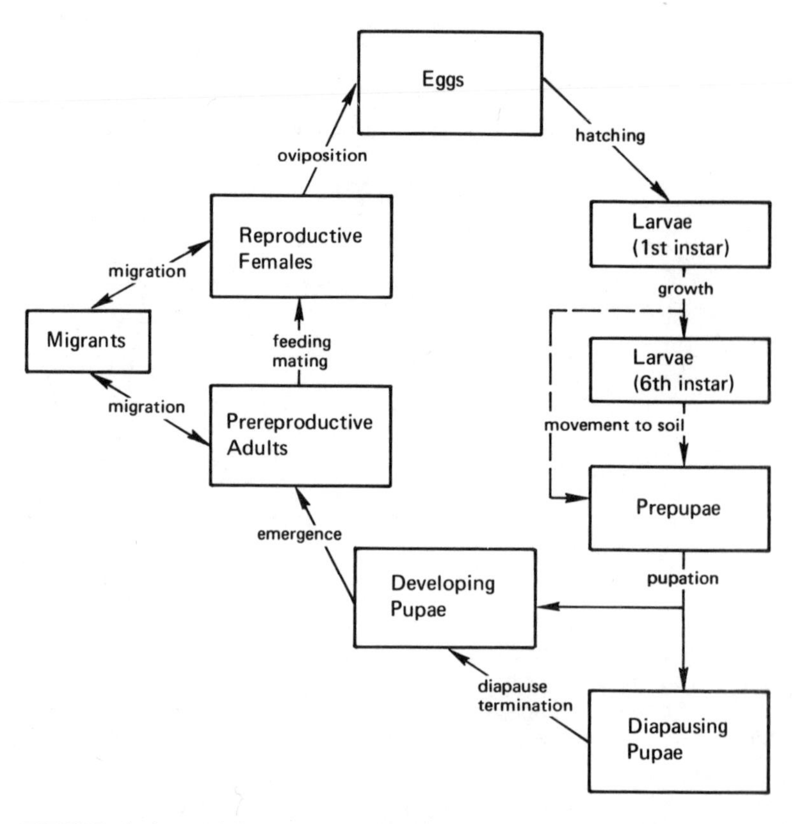

FIGURE 4-2. Life system flow diagram for *Heliothis zea*. (Source: Stinner, personal communication.)

The principal cotton plant model, SIMCOT II, was developed at Mississippi State University. SIMCOT II is a dynamic simulator that describes the growth of a cotton plant from emergence to open boll. Exogenous variables are minimum and maximum temperatures, rainfall, solar radiation, pan evaporation, and nitrogen level. The simulator prints out a daily plant map which shows the position and condition of each fruiting site on the plant. At the end of a simulated season, the yield is calculated. SIMCOT II accurately predicts observed cotton plant growth if nitrogen and water are not limiting factors. SIMCOT II has also been used to assess the effects of fertilizer levels, and the time and intensity of insect damage upon a cotton crop (McLaughlin 1973). Simulations based upon 1972 field data predicted yields within 8 percent of the observed yields. The use of simulation models is also important in the development of plant varieties that are resistant to pests. An example is the application of SIMCOT II to a problem with an experimental strain of cotton with the mutant frego-bract (Jenkins 1973).*

Perhaps the most important use of mathematical models in the development of biological, cultural, or integrated control programs, is to organize and coordinate data from the experiments on different parts of the ecosystem. A systems analysis approach clarifies what additional information must be collected before making recommendations based on predictions of whole system response.

REFERENCES

Adkisson, P. L. (1962) Timing defoliants and desiccants to reduce over-wintering populations of the pink bollworm. J. Econ. Entomol. 55:949-951.

Adkisson, P. L. (1964) Action of the photoperiod in controlling insect diapause. Am. Nat. XCVIII (902):357-374.

Adkisson, P. L. (1966) Internal clocks and insect diapause. Science 154:234-241.

Adkisson, P. L. (1969) How insects damage crops. Pages 155-164 *in* How crops grow--A century later. Conn. Storrs Agric. Exp. Stn. Bull. 708.

Adkisson, P. L. (1972) The integrated control of the insect pest of cotton. Proc. Tall Timbers Conf. Ecol. Animal Control by Habitat Management. 4:175-188.

---

*Frego-bract cotton is quite resistant to boll weevil damage. However, in some locations frego-bract cotton matures slowly and produces reduced yields. Jenkins considered three possible causes of this delayed maturity: early season bug damage to plants; squares that are overly sensitive to solar radiation; or a longer time period required between squaring. The results of the SIMCOT II simulation of each of these situations suggested that early season bug damage to plants is causing the problem. This hypothesis must, of course, be tested further in the field, but the results of the simulation can be used to develop a well designed field experiment which will have a high probability of success. If in fact early plant bug feeding is shown to be the cause, the problem may be controlled by an early insecticide treatment. The total amount of insecticide used in such a program would be considerably less than currently used amounts.

Adkisson, P. L. (1974) Alternatives to the unilateral use of insecticides for insect pest control in certain field crops. Proc. Symp. Ecol. and Agric. Prod. Univ. Tenn. In press.

Adkisson, P. L., and J. C. Gaines (1960) Pink bollworm control as related to the total cotton insect control program of central Texas. Tex. Agric. Exp. Stn. Misc. Publ. 444. 7 pp.

Adkisson, P. L., L. H. Wilkes, and B. J. Cochran (1960) Stalk destruction and plowing as measures for controlling the pink bollworm. J. Econ. Entomol. 53:791-794.

Adkisson, P. L., R. A. Bell, and S. G. Wellso (1963) Environmental factors controlling the induction of diapause in the pink bollworm. J. Inst. Physiol. 9:299-310.

Adkisson, P. L., D. R. Rummel, W. L. Sterling, and W. L. Owen, Jr. (1966) Diapause boll weevil control: a comparison of two methods. Tex. Agric. Exp. Stn. Bull. 1054.

Annand, P. N. (1948) Recent developments in the control of cotton insects. A report to the Cotton Subcommittee of the House Committee on Agriculture. U.S. Dep. Agric. Bur. Entomol. Pl. Quar. p. 1021-1042.

Bottrell, D. G. (1972) New strategies for management of the boll weevil. Proc. 5th Ann. Tex. Conf. Insect, Plant Disease, Weed Brush Control. Tex. A & M Univ. p. 67-72.

Brazzel, J. R. (1961) Boll weevil resistance to insecticides in Texas in 1960. Tex. Agric. Exp. Stn. Prog. Rep. 2171. 4 p.

Brazzel, J. R., H. Chambers, and P. J. Hammon (1961) A laboratory rearing method and dosage-mortality data on the bollworm, *Heliothis zea*. J. Econ. Entomol. 54: 949-952.

Cross, W. H. (1973) Biology, control and eradication of the boll weevil. Annu. Rev. Entomol. 18:17-46.

Cross, W. H., M. J. Lukefahr, P. A. Fryxell, and H. R. Burke (1975) Host plants of the boll weevil. Environ. Entomol. 4:19-26.

Dunnam, E. W., and J. C. Clark (1941) Cotton aphid multiplication following treatment with calcium arsenate. J. Econ. Entomol. 34:587-588.

Eden, W. G., H. C. Chiang, E. H. Glass, D. L. Haynes, P. Oman, and H. T. Reynolds (1973) The pilot boll weevil eradication experiment. Bull. Entomol. Soc. Amer. 19:218-221.

Ewing, K. P., and E. E. Ivy (1943) Some factors influencing bollworm populations and damage. J. Econ. Entomol. 36:602-606.

Falcon, L. A. (1971) Microbial control as a tool in integrated control programs. Pages 346-364 *in* C. B. Huffaker, ed. Biological control. Plenum Press, New York.

Falcon, L. A., R. van den Bosch, C. A. Ferris, L. K. Stromberg, L. K. Etzel, R. E. Stinner, and T. F. Leigh (1968) A comparison of season-long cotton pest-control programs in California during 1966. J. Econ. Entomol. 61:633-642.

Fenton, F. A., and W. L. Owen (1953) The pink bollworm of cotton in Texas. Tex. Agric. Exp. Stn. Misc. Publ. 100.

Folsom, J. W. (1928) Calcium arsenate as a cause of aphid infestation. J. Econ. Entomol. 21:174.

Gorham, D. P. (1847) The cotton worm: DeBow's Commercial Rev. 3:535.

Hagen, K. S., R. van den Bosch, and D. L. Dahlsten (1971) The importance of naturally occurring biological control in the western United States. Pages 253-293 *in* C. B. Huffaker. Biological control. Plenum Press, New York.

Hardee, D. D., W. H. Cross, and E. B. Mitchell (1969) Male boll weevils are more attractive than cotton plants to female boll weevils. J. Econ. Entomol. 62:165-169.

Hunter, W. D. (1912) The control of the boll weevil. U.S. Dep. Agric. Farmer's Bull. 500. 14 p.

Ignoffo, C. M. (1970) Microbial insecticides: no-yes; now-when! Proc. Tall Timbers Conf. Ecol. Animal Control by Habitat Management. 2:41-58.

Isely, D. (1928) The relation of leaf color and leaf size to boll weevil infestation. J. Econ. Entomol. 21:553-559.

Isely, D. (1934) Early varieties of cotton and boll weevil injury. J. Econ. Entomol. 27:762-766.

Isely, D. (1948) Methods of insect control. Part I, 3rd ed. Burgess Publ. Co., Minneapolis, Minn. 134 p.

Jenkins, J. A. (1973) Systems analysis and the evaluation of morphogenic characters in cotton. Miss. State Univ. Symp. Application of Systems Methods to Crop Production.

Jenkins, J. N., and W. L. Parrott (1971) Effectiveness of frego bract as a boll weevil resistance character in cotton. Crop Sci. 11:159.

Keller, J. C., E. B. Mitchell, G. McKibben, and T. B. Davich (1964) A sex attractant for female boll weevils from males. J. Econ. Entomol. 57:609-610.

Klassen, W., E. F. Knipling, and J. U. McGuire (1970) The potential for insect population suppression by dominant conditional lethal traits. Ann. Ent. Soc. Amer. 63:238-255.

Knipling, E. F. (1966) Some basic principles in insect population suppression. Bull. Entomol. Soc. Amer. 12: 7-15.

Knipling, E. F. (1971) Boll weevil and pink bollworm eradication: progress and plans. Ginner's J. Yearbook 1971:23-32.

Knipling, E. F. (1972a) Entomology and the management of man's environment. 14th Intern. Congr. Ent. Report. pp. 4-18.

Knipling, E. F. (1972b) Sterilization and other genetic techniques. *In* Pest Control Strategies for the Future. Washington, D.C.: National Academy of Sciences, pp. 272-287.

Lindgren, P. D. (1969) Approaches to the management of *Heliothis* spp. in cotton with *Trichogramma* spp. Proc. Tall Timbers Conf. Ecol. Animal Control by Habitat Management 1:207-218.

Lloyd, E. P., W. P. Scott, K. K. Shaunak, F. C. Tingle, and T. B. Davich (1972) A modified trapping system for suppressing low density populations of overwintering boll weevils. J. Econ. Entomol. 65:1144-1147.

Lukefahr, M. J., L. W. Noble, and D. F. Martin (1964) Factors inducing diapause in the pink bollworm. U.S. Dep. Agric. Tech. Bull. 1304.

Lukefahr, M. J., C. B. Cowan, and J. E. Houghtaling (1970) Field evaluations of improved cotton strains resistant to the cotton fleahopper. J. Econ. Entomol. 63:1101-1103.

Maley, F. W. (1902) Report on the boll worm. Tex. A & M Univ. College Station. 45 p.

Maxwell, F. G., J. W. Jenkins, and W. L. Parrott (1972) Resistance of plants to insects. Adv. Agron. 24:187-265.

McLaughlin, R. E. (1973) The economic threshold and the interface between plant and insect models. Miss. State Univ. Symp. Application of Systems Methods to Crop Production.

Mueller, A. (1971) Bionomics of lygus bugs in the safflower-cotton-alfalfa seed cropping system in California. Ph.D. Diss. Dep. Entomol. Univ. Calif. Riverside. 141 pp.

N.A.S. (1969) Insect pest management and control. 3. Washington, D.C. 508 p.

Newsom, L. D. (1970) The end of an era and future prospects for insect control. Proc. Tall Timbers Conf. Ecol. Animal Control by Habitat Management. 2:117-136.

Newsom, L. D., and J. R. Brazzel (1968) Pests and their control. Pages 367-405 *in* F. C. Elliot, M. Hoover, and W. K. Porter, Jr., eds. Advances in production and utilization of quality cotton: principles and practices. Iowa State University Press, Ames.

Noble, L. W. (1969) Fifty years of research on the pink bollworm in the United States. U.S. Dep. Agric. Handbook 357. 62 p.

Ohlendorf, W. (1926) Studies of the pink bollworm in Mexico. U.S. Dep. Agric. Bull. 1374.

Pierce, W. D. (1908) Studies of parasites of the cotton boll weevil. U.S. Dep. Agric. Bur. Entomol. Bull. 73.

Pierce, W. D., R. A. Cushman, and C. E. Hood (1912) The insect enemies of the cotton boll weevil. U.S. Dep. Agric. Bur. Entomol. Bull. 100.

Reynolds, H. T., P. L. Adkisson, and R. F. Smith (1975) Insect pest management in cotton. *In* R. L. Metcalf and W. H. Luckman, eds. Introduction to pest management. John Wiley and Sons, New York.

Ridgway, R. L. (1969) Control of the bollworm and tobacco budworm through conservation and augmentation of predaceous insects. Proc. Tall Timbers Conf. Ecol. Animal Control by Habitat Management. 1:127-144.

Riley, C. V. (1873) Fifth annual report on the noxious, beneficial and other insects of the state of Missouri, made to the State Board of Agriculture, pursuant to an appropriation for this purpose from the Legislature of the state. 8th Annu. Rep. State Board Agric. 1872. p. 160-168.

Robertson, O. T., V. L. Stredonsky, and D. H. Currie (1959) Kill of pink bollworms in the cotton gin and the oil mill. U.S. Dep. Agric. Prod. Res. Rep. 26. 22 p.

Schuster, M. F., and F. G. Maxwell (1974) The impact of nectariless cotton on plant bugs, bollworms, and beneficial insects. Proc. 1974 Beltwide Cotton Res. Conf.

Smith, R. F., and R. C. von Borstel (1972) Genetic control of insect populations. Science 178:1164-1174.

Smith, R. F., and L. A. Falcon (1973) Insect control for cotton in California. Cotton Grow. Rev. 50:15-27.

Sterling, W. L., and R. L. Haney (1973) Cotton yields climb, costs drop through pest management systems. Tex. Agric. Exp. Stn. Prog. Rep. 19:4-7.

Stern, V. M. (1966) Significance of the economic threshold in integrated pest control. Proc. FAO Symp. Integrated Pest Control. p. 2, 41-56.

Stern, V. M. (1969) Interplanting alfalfa in cotton to control lygus bugs and other insect pests. Proc. Tall Timbers Conf. Ecol. Animal Control by Habitat Management. 1:55-69.

Stern, V. M., R. van den Bosch, T. F. Leigh, O. D. McCutcheon, W. R. Sallee, C. E. Houston, and

M. J. Garber (1967) Lygus control by strip-cutting alfalfa. Univ. Calif. Agric. AXT-241. 13 pp.

Stern, V. M., P. L. Adkisson, O. G. Beingolea, and G. A. Viktorov (1975) Cultural controls. *In* C. B. Huffaker, ed. Theory and practice of biological control. (in press).

Stinner, R. E., R. L. Rabb, and J. R. Bradley (1974) Population dynamics of *Heliothis zea* (Boddie) and *H. virescens* (F.) in North Carolina: a simulation model. Environ. Entomol. 3(1):163-168.

van den Bosch, R., T. F. Leigh, L. A. Falcon, V. M. Stern, D. Gonzales, and K. S. Hagen (1971) The developing program of integrated control of cotton pests in California. Pages 377-394 *in* C. Huffaker, ed. Biological Control. Plenum Press, New York.

Walker, J. K., Jr., and J. A. Niles (1971) Population dynamics of the boll weevil and modified cotton types. Tex. Agric. Exp. Stn. Bull. 1109. 14 p.

Walker, J. K., Jr., J. V. Robinson, J. A. Niles, J. R. Gannoway, and C. F. Muska (1972) Cotton fleahopper infestations and damage in different cotton genotypes. Tex. Agric. Exp. Stn. Dep. Entomol. Tech. Rep. 21.

Whitcomb, W. H. (1970) History of integrated control as practiced in the cotton fields of the south central United States. Proc. Tall Timbers Conf. Ecol. Animal Control by Habitat Management. 2:147-155.

Whitcomb, W. H., and K. Bell (1964) Predaceous insects, spiders, and mites of Arkansas cotton fields. Arkansas Agric. Exp. Stn. Bull. 690.

Whitten, M. J., and G. G. Foster (1975) Genetical methods of pest control. Annu. Rev. Entomol. 20:461-476.

Wilkes, L. H., P. L. Adkisson, and B. J. Cochran (1959) Stalk shredder tests for pink bollworm control. Tex. Agric. Exp. Stn. Prog. Rep. 2095.

5

DEVELOPMENTS IN THE COTTON INDUSTRY WITH POTENTIAL IMPACT ON PEST MANAGEMENT

Many of the developments taking place within the cotton industry are peripheral to pest management problems. However, some of these developments could have an impact on future pest control practices.

COTTONSEED PROTEINS IN HUMAN NUTRITION

The manufacture of high-quality food protein from cotton-seeds represents a technological advance that may alter the economic future of the cotton industry. The first commercial plant to produce human food grade protein flour from cottonseed is now in operation. The plant is owned and operated by the Plains Cooperative Oil Mill of Lubbock, Texas, and is designed to produce about 50,000 lb of such flour per day. The introduction of cottonseed protein products to the human food market could greatly increase the dollar value of the annual cotton crop. Consequently, this could change the economic basis on which agricultural management decisions are made, including those decisions related to pest management.

With each 100 lb of fiber, cotton plants produce about 170 lb of seed, of which only 5 percent is needed for planting. Seed production averages about 850 lb per acre, but the seed market value has been only 10 to 15 percent of the total value of the crop. Refined cottonseed oil is used commercially in human food products. The hulls and meals (oil-extracted residue) are used primarily in livestock feeds. Because of the presence of gossypol, hulls and meal are toxic or deleterious to nonruminants. Gossypol and related pigments also increase the cost of refining the cottonseed oil, reducing its competitiveness with other plant oils.

Cottonseed meal is more than 50 percent protein and, were it not for the gossypol, is nutritionally adequate for human consumption. The world protein shortage is expected to rise steadily, reaching a level of 2,157,000 metric tons by 1975. In 1971, the major cotton-growing countries produced cottonseed containing 4,700,000 metric tons of high-quality protein. Only a very small portion could be used for human nutrition. Were these quantities rendered fit for human consumption, the annual production of cottonseed protein could greatly reduce the projected protein shortage.

High-quality edible proteins can now be produced from cottonseed. Previously restricted to low quantity production from seeds of glandless cotton varieties (gossypol-free), edible protein production can now use ordinary gland varieties. The development of technology for food-protein production from gland cottonseeds is far superior to relying on the limited possibilities for protein manufacture from cottonseed produced by glandless varieties. Gossypol-free (glandless) varieties of cotton are grown on a limited experimental basis and have displayed some serious pest control problems. Gossypol is apparently important in the defense of the cotton plant against certain insect species; its absence in the glandless varieties renders the plant relatively defenseless.

A commercially feasible process for separation of glands containing gossypol from cottonseed was the key development leading to the production of food grade protein. In cost, nutritional value, and physical properties, the cottonseed protein products are fully competitive with soybean protein products. In addition, cottonseed protein has the advantage of being more bland than soybean protein, so its potential uses in food manufacture may be greater than those of soybean products. The production of food grade protein from cottonseed meal does not reduce the industry's capacity for cottonseed oil production or the use of hulls as feed supplements for livestock.

In the past, cottonseed has been of low market value, the return to the grower being little more than sufficient to pay the cost of ginning. With the advent of protein production for human consumption, the market value of cottonseed may be predicted to rise markedly over the next decade. This predicted upward trend is already beginning; cottonseed price at the gin yard was $50 per ton in August 1972 but had risen to $100 per ton by August 1973. The prediction is also based on the generality that a crop grown for food returns more to the producer than the same crop grown for animal feed. Food grade soybeans and peanuts sell for two to three times the price obtained for the same products used for animal feed.

HIGH-DENSITY PLANTINGS

Although cotton plant populations per acre have been increasing for some time, the potential role of high-density planting in controlling certain insect and disease pests has been held back by the need for effective herbicides and harvesting machinery. High-density planting depends on short, uniform, early-maturing cotton plants. It has potential for increasing or maintaining yields while lowering production costs and shortening the production period. The shorter growing period can help control some cotton insects and diseases.* However, the practice requires effective weed control. Because narrow-row culture or broadcast seeding does not allow between-row cultivation, herbicides are required. Presently available herbicides are being used in high-density plantings. However, this increases potential for ecological shifts in weed species. Thus, the use of high-density plantings to shorten the season and reduce insecticide or fungicide application could result in need for an increased variety of herbicides.

GROWTH REGULATORS

Other than harvest aid chemicals (defoliants and desiccants), growth regulators have no demonstrated commercial use in cotton production. However, there appears to be increased interest in these materials for use on large acreage crops.† In cotton production, growth regulators might be directed toward establishing more uniform plants, compressing the fruiting and harvesting periods, and avoiding late-season insect pests.

---

*For example, early-maturing plants may partially escape the effects of *Phymatotrichum* root rot. Also, *Verticillium* wilt does not appear to occur on as high a percentage of plants in high-density stands.

†In 1973, commercial use of a growth regulator in many peanut producing areas of the South seemed to inhibit the growth of peanut foliage without reducing root development of peg and nut production. This reduced insect and disease control problems.

Compounds that improve seed and seedling performance, particularly against environmental stress, could be helpful in disease control. Growth habit and flowering patterns of cotton might be modified. Producing more compact plants and lowering fruiting areas could lessen exposure to the environment and increase efficiency of pesticide application. Regulators might be used to increase fruit set in areas subject to late season stress. A growth regulator that would terminate flower production but allow full development of fruit already present would help solve many critical insect problems.*

NAKED SEED VARIETIES

After separation from the lint fibers in the ginning process, the seed of commercially grown American upland cotton has a dense short fuzz covering the entire seed coat. The seed must be reginned or delinted prior to processing by cottonseed oil mills. These delinting and other seed handling operations comprise a significant portion (perhaps as much as half) of the oil mill processing costs. The other use of cottonseed, planting, requires that most or all of the short fuzz covering be removed in order for planting machinery to operate properly. Planting seed may be delinted by machine or acid. Both processes are costly and potentially damaging to the seed.

Naked seed varieties lack the short fuzz covering on the seed coat. Because these varieties would not require delinting, their potential for reducing processing and seed costs appears to be substantial. In addition, naked seed varieties may play an indirect role in reducing the need for late season weed control and reducing the quantities of defoliants and desiccants used in harvesting. Machine harvested cotton, whether harvested by spindle-picker or stripper, contains leaf and weed trash. This trash has to be removed in the ginning process. Green plant matter can stain the lint and lower the grade. Much of the problem might be alleviated if the lint could be

---

*An apparent increase in *Verticillium* wilt resistance has been shown experimentally by the use of certain growth retardants. The action is thought to be allied with induction of naturally occurring biochemicals (phytoalexins) in the plant. (Buchenour and Erwin 1973a, 1973b, Zaki et al. 1972a, 1972b).

stored and the trash allowed to dry before processing. However, the storage of seed cotton usually generates enough heat to damage the seed.

The commercial planting of naked seed varieties would allow the development of cotton harvesting machinery which separates the seed from the lint in the field. This process could allow the lint to be stored and dried without damaging the seed. This, in turn, might reduce the need for late season weed control and the amount of defoliants and desiccants required as harvest aid chemicals.

HARVESTING AND GINNING DEVELOPMENTS

Harvesting Methods

Cotton planted in narrow rows or broadcast may fruit and mature as much as one month ahead of normal cottons with no yield reduction. In addition, short season or fast fruiting varieties of cotton, which produce reduced yield when planted in standard rows, are made competitive when planted in rows one-half meter apart. The development of harvesting techniques allowing cotton to be planted and harvested in narrow rows or broadcast may be an advantage in boll weevil and bollworm control programs. The early maturing characteristic of this type of planting could allow the plant to "outrun" these cotton insects.

Cotton is machine harvested by either spindle pickers or strippers. Multiple harvests of the same field can be made with a picker. The stripper is a once-over harvester. The method of harvest planned usually determines the variety to be planted. Geographically, stripper harvesters are used in Oklahoma, most of Texas, and parts of New Mexico. Relatively low yields are characteristic of these areas and the stripper substantially lowers harvest costs.

Where strippers are used, growers may kill the cotton plants with a desiccant when at least 90 percent of the bolls have opened. Alternatively, they may delay harvest until frost has killed the plants. (If yields are expected to be less than one-half bale per acre, harvest is often delayed.) If the crop is stripper harvested without being killed, the mass of green leaves and immature bolls cannot be ginned. Some form of turnrow storage might allow the leaf trash to dry and the immature bolls to become ginnable. However, such storage would generate sufficient heat to damage the seed.

New harvesting innovations include the experimental development of cotton combines. In combining, the cotton plants are cut off and borne through the picking machinery (spindles) in a once-over harvest operation. (Standard machines move picking spindles in and out of the standing plants as the machine passes. Use of these standard machines permits several successive harvests of the same plants.) The cotton combine was developed to harvest cotton planted in rows less than one meter apart or in no rows (broadcast). Prototype stripper harvesters capable of harvesting cotton planted in 1/2-meter rows were built and field-tested successfully in Arkansas in 1973. No commercial machines using brushes to strip rows closer than approximately 80 cm have been developed. However, finger-type strippers capable of harvesting cotton planted in any or no row pattern have been in use more than five years.

Storage and Handling Methods

The development of new storage and handling techniques may play an important role in determining the feasibility of integrated pest management systems. Large regions of the Cotton Belt can be planted in a short period of time. Consequently, large portions of the crop can be ready for harvest at the same time. Often, it is impossible for cotton gins in an area to process the maturing crop as rapidly as it can be harvested. Prolonging the cotton harvest, because ginning capacities are not balanced with crop maturity and harvest rates, could undermine any insect pest management system requiring early stalk destruction.

It is feasible to store harvested seedcotton for indefinite periods before ginning, if seed damaging heat buildup is prevented and suitable types of machinery or storage facilities are available. Prototype machines were designed and developed by 1969 to move seedcotton in and out of storage. By 1972, commercial machinery designed to build holding stacks of unginned cotton and to deliver the cotton to gins were given extensive field tests. It appears that there are gains in quality of fiber, associated with the earlier harvest completion and the "curing" process in the stored seedcotton.

Ginning Techniques

Between the farm and the textile mill door, the largest cost is the ginning process. This involves separating the seed from the lint and packaging the lint into 500-lb bales. The increasing demand for ginning capacity has resulted in larger gins operating for shorter periods of time during the crop year. The development of storage and handling techniques may serve to lengthen the ginning season and extend use of present ginning capacity. Another concept involves the development of a low energy gin to operate in the field with the picking operation. Some research has been started in this area. If naked seed varieties were available, it seems probable that simple ginning and packaging devices could be designed for mounting on standard mechanical harvesters.

PROCESSING AND UTILIZATION TECHNIQUES IN THE TEXTILE INDUSTRY

Unlike other agricultural products, cotton fibers go to end uses in which competition is almost exclusively with industrial products. These are the synthetic fibers. Cotton has faced and continues to face a declining share of the fiber market. There appear to be three major reasons for cotton's declining role: price competition, supply reliability, and fiber characteristics. While synthetic fiber prices have been declining in recent years, cotton prices have fluctuated widely. Cotton supplies, including supplies of different staple lengths, vary considerably from year to year. Synthetic fiber production is not subject to varying weather, pests, and so on. Perhaps more important, synthetic fiber production takes place year round. Textile manufacturers do not have to maintain as large inventories of synthetic fibers as of cotton. Finally, man-made fibers are waste-free and more easily made to uniform staple length. Thus, although the textile technology was built around the processing of cotton, crop consumption trends have been downward.

However, there appear to be a number of developments within the textile industry with a potentially significant impact on cotton production. While these developments affecting the utilization of cotton may not increase consumption of the crop, they may encourage change in the types of cotton produced commercially. This, in turn, may alter important production practices, especially pest control.

Production of cotton-synthetic fiber blends is increasing rapidly. Typically, these blends are "easy care" (durable press) fabrics in a 50:50 or 65:35 ratio of polyester fiber to cotton. Improved technology has allowed the use of shorter staple cottons in good quality blends than in all cotton fabrics of similar quality. Increasing demand for blends could alter the demand for cotton staple lengths.* Although cotton fiber strength is not a major factor in fabric blends, the cotton in blends should have a fiber length distribution similar to that of the synthetic fiber and the cotton must be free of foreign matter. Thus, an increased demand for cotton in fabric blends might encourage the adoption of varieties having the desired fiber characteristics. These varieties might incorporate other plant characteristics, such as early maturity or naked seed, with implications for pest control and other production and harvesting practices.

Another development with implications for cotton production is the potential adoption of open-end spinning. Conventional ring spinning has continually increased processing speeds. The increased operating speeds have placed additional stress on cotton fibers, requiring stronger fibers and longer staples. However, the upper limits of speed in operating ring spindle machinery may have been reached and it is conceivable that textile manufacturers may turn to the faster operating open-end spindles for the yarnmaking process. Although only a small fraction of the total spindles now in operation are open-end, an increase in their use could significantly change the quality of cotton demanded. Open-end spindles can utilize short staple fibers, but require high fiber uniformity and less trash.

A third potential development is the use of cotton in nonwoven (disposable) fabrics. However, this end-use requires the absence of trash in the raw stock. If short and coarse fiber cotton could be cleaned to an acceptable degree, the potential for types of cotton in the nonwoven market could be significant.

These potential developments in the end-use of cotton fibers might provide an opportunity to use new varietal types in production schemes designed to improve pest management practices. In the past, processing requirements prevented the commercial development of short season

---

*Historically, cotton from a given production area has become associated with certain quality factors. Cottons from the West are assumed to possess high breaking strength and relatively long staples. West Texas and Oklahoma traditionally have produced short-staple, low-grade cotton. Delta cotton is characterized by good strength and good staple length.

varietal types. The agronomic and harvesting practices made possible by new varietal types demanded by the textile industry may become an integral part of future pest management systems.

REFERENCES

Buchenauer, H., and D. C. Erwin (1973a) Effect of root and foliage treatment with a peperazine derivative fungicide on Verticillium (albo-atrum) wilt of cotton. Plant Dis. Rep. 57:460-462.

Buchenauer, H., and D. C. Erwin (1973b) Systemic fungicidal effect of thiophamate methyl on Verticillium (albo-atrum) wilt of cotton and its transformation to methyl 2-benzimidiazole-carbamate in cotton plants. Phytopathology 63:1091-1095.

Zaki, A. I., N. T. Keen, J. J. Sims, and D. C. Erwin (1972a) Vergosin and hemigossypol, antifungal compounds produced in cotton plants inoculated with Verticillium albo-atrum. Phytopathology 62:1398-1401.

Zaki, A. I., N. T. Keen, J. J. Sims, and D. C. Erwin (1972b) Implication of vergosin and hemigossypol in resistance of cotton to Verticillium albo-atrum. Phytopathology 62:1402-1406.

6

DELIVERY OF INTEGRATED PEST MANAGEMENT PROGRAMS

The use of improved knowledge of cotton ecosystems offers the greatest potential for better pest management in the future. New material inputs--selective pesticides, biological controls, and resistant plant species--are foreseeable. Yet, no single material or any combination of materials will "eliminate" the need for intensive crop management. Knowledge of the agroecosystem and how it responds to natural and introduced variables is the key to reducing both pest losses and environmental contamination. Existing knowledge and knowledge currently being generated will not be of any value until it is delivered to and implemented on the farm.

The introduction of new practices to agriculture is almost always more difficult than the introduction of new materials. Material inputs can be produced in great quantities by a few manufacturers, distributed through existing marketing establishments, and utilized by the farmer in much the same manner as previous materials were used. At each step--production, distribution, and final use--the profit motive speeds the process of adoption within an existing network of interdependencies. Knowledge differs from material inputs. Because knowledge cannot be "owned," it is difficult to profit from its production and dissemination. Ironically, because knowledge dissemination cannot be protected by an exclusive copyright, the private market sector of our economy produces and distributes knowledge inadequately. The utilization of new knowledge is also different. The receiver must be sufficiently intelligent and educated to understand the message; he must integrate it with other management principles and the characteristics of his farm; and he must change his management techniques. These differences indicate a government role in the production, distribution, and utilization of knowledge. This section of the report considers all three aspects of the problem for cotton pest management knowledge, with an emphasis on the factors which most closely relate to delivery systems.

LEGAL REQUIREMENTS

The success of integrated insect control programs depends on the presence of a vigorous pest management industry and collective institutions to ensure that the management programs which are implemented will be effective. This section will examine the major legal problems which are likely to arise in the development of pest management delivery systems and effective collective institutions.

Licensing and Liability

The role of the law in the creation of a pest management industry is basically twofold: incentives must be balanced against constraints. It is assumed that the industry is in the public interest, and the first role of the law is to encourage entry to the field by creating incentives or at least removing entry barriers. However, since every new industry potentially imposes costs to the users of its products and services, as well as to third parties and future generations, society has an interest in minimizing external costs associated with the activity. This interest can be represented by the standards of liability to which the industry is held. However, consideration should be given to regulatory mechanisms which assess possible environmental and other side effects of integrated control.

An industry which performs services can be encouraged by laws which accord it recognition and induce public confidence by establishing qualifications for the provision of its services. Such recognition is often accorded by licensing the activity. The basic legal justification for licensing is the desirability of establishing a minimum level of services affecting the public by prescribing standards of practice. Another equally important reason for licensing is industry desire to increase its status. A less meritorious legal justification for licensing is control of entry to the field. Pest control advisors are now licensed in California (West Cal. Ag. Code 11402 et seq.). The principal licensing and license revocation standards are limited to requirements related to personnel competence and to mandatory disclosure to the public of the risks the service entails. The legislation is consistent with the most restrictive police power justification for occupational licensing:

> Where the practice of a profession or calling requires special knowledge or skill and intimately affects the public health, morals, order or safety, or the general welfare, the legislature may prescribe reasonable qualifications for persons desiring to pursue such profession or calling and require them to demonstrate their possession of such qualifications by examination . . .(*State* v. *Ballance* 1949).

The key sections of the California legislation illustrate the standards drafted to implement these objectives:

> §12022.
> Applicants for licensing shall elect to be examined for certification in one or more of the following categories:
> (a) Control of insects, mites, and other invertebrates.
> (b) Control of plant pathogens.
> (c) Control of nematodes.
> (d) Control of vertebrate pests.
> (e) Control of weeds.
> (f) Defoliation.
> (g) Plant growth regulation.
> The examination shall be prepared and administered by the director.
>
> §12023.
> An agricultural pest control adviser license may be refused or may be revoked or suspended by the director as necessary to carry out the purposes of this division. Cause for refusal, revocation, or suspension shall include, but shall not be limited to, any of the following:
> (a) Failure to put a recommendation in writing.
> (b) The making of false or fraudulent statements in any written recommendation.
> (c) Failure or refusal to comply with any provisions of this chapter, or any other regulation adopted by the agricultural commissioner.
> (d) Failure or refusal to comply with any provisions of this division or of Division 7 (commencing with Section 12501) relating to pesticides or regulation of the department adopted pursuant to such provisions.

(e) Failure to qualify by examination in at least one of the categories in Section 12022.

§12024.

The director by regulation shall establish the minimum requirement for pest control adviser by education and examination to carry out the purposes of this division. The director may, after examination, issue a provisional license to an applicant, pending his certification pursuant to Section 12022, provided that no such provisional license shall be issued or effective after January 1, 1974. The director shall by regulation establish equivalent experience qualifications in lieu of education and examination for a provisional license as a pest control adviser up to January 1, 1974; therefore, he may establish such qualifications if he determines that experience qualifications are necessary to carry out the purposes of this chapter.

The question may arise whether pest management should be considered a profession. All licensed activities are sometimes loosely referred to as professions, but a more restricted definition of profession is traditionally used by sociologists and, to some extent, by the law. A profession is generally defined in terms of association with a body of theoretical knowledge and a service orientation which is free from the constraints of the client or the state to define an acceptable work product (Mills and Tothman 1968). State regulation generally delegates authority to the professional organization to regulate entry and to establish the standards of practice. The legal significance of classifying an activity as a profession under this standard is that the courts are more likely to accept internal professional practices as standards defining conduct subject to liability. Members of nonprofessional occupations are more likely to be judged by standards of care external to the occupations. A professional offers a skill and the standard by which this skill is judged for the purposes of imposing financial liability for losses suffered by the users is "the general average of professionally acceptable conduct" (Curran 1959). The technical significance of classifying an activity as a profession under the restricted definition is that breach of duty must be established by expert testimony as in the case of medical malpractice. The necessity of using experts to determine if conduct should be subject to liability does not, of course, make the activity a profession.

Pest management is hard to classify. It rests on a body of theoretical knowledge, but it is difficult to argue that it has a service orientation differing from that of any other service consumed by the public. The relationship between manager and client is less one of trust and confidence than contract. It could be argued, however, that it is analogous to medical treatment. On balance, there seems to be no compelling reason to classify pest management as a profession for purposes of liability determination. It has been argued that the medical profession, for example, should be held to lower standards of liability to encourage medical practice, but there seems to be no reason to hold pest management advisors to standards lower than those applied to any other class of services offered to the public.

The law imposes liability for losses suffered as a result of an activity on two grounds: fault and strict liability or nonfault. Liability based on fault is imposed if an actor intentionally causes harm or fails to exercise reasonable care toward a person to whom a duty is owed, e.g., is negligent. Strict liability is imposed on several grounds. It was originally imposed by common law for the maintenance of an ultrahazardous activity. An ultrahazardous activity is generally one which is abnormal for the particular locale and one which a court finds likely to result in harm "from that which makes the activity ultrahazardous, although the utmost care is to prevent the harm" (Restatement of Torts, Section 519). Increasingly, liability has been imposed on manufacturers for products with defective conditions which are unreasonably dangerous to the consumer [Restatement of Torts, Second Section 402(A)]. In addition, strict liability has been imposed by the courts under the sales concept of implied warranty. Warranties may be imposed either under the Uniform Commercial Code (UCC), in force in all states except Louisiana, or by the courts. It is unlikely that the UCC would apply to many activities of pest management consultants because it pertains only to goods and not services. However, the line between transactions involving the sale of goods and the sale of knowledge is not clear and has been characterized as "merely a verbal formula in which results are expressed" for a variety of factors, such as the need to encourage the activity (White and Summers 1972). However, warranties may be imposed on nonsale of goods transactions by the courts, for it is generally agreed that the UCC is not intended to stifle the growth of implied warranties of merchantability, and, thus, the courts are free to draw analogies from the UCC to transactions not covered by it. Under both tort and sales theories, an actor is liable regardless of fault, but the defenses available to the defendant differ under each theory. If liability is imposed on tort grounds, the manufacturer cannot avoid liability through the use of disclaimers, but can avoid liability if the user assumes the risk. If, however, liability is imposed on warranty theories, liability may be avoided by properly drafted disclaimers.

In three states, Louisiana, Oklahoma, and Oregon, liability for damage caused by spraying crops on adjoining land has been classified as ultrahazardous. The courts have focused on the danger inherent in the practice of spray application regardless of the fact that spraying is a normal agricultural practice (*Gotreaux* v. *Gary* 1957, *Loe* v. *Lenhardt* 1961, *Young* v. *Darter* 1961). In other jurisdictions liability for injuries resulting from pesticide use is imposed only if negligence is shown.

Pesticide manufacturers have been held liable for negligence when an insecticide applied according to instructions caused damage to a crop as well as to the target pest species on the grounds the pesticide was misbranded. The failure to warn about the effect on nontarget crops was held negligent and the court held that the pesticide was misbranded. Thus, the manufacturer was liable on a theory of negligence per se (*Perry Creek Cranberry Corporation* v. *Hopkins Agricultural Chemical Co.* 1966). Spraying in wind which causes the pesticide to drift onto the land of a third party is a frequent basis for liability (*Leonard* v. *Abbot* 1963).

The doctrine of *res ipsa loquitar* has been applied to allow cotton growers to recover damages when 2,4-D and Silvex were present in toxaphene sprayed onto a field. The inference that the defendant was negligent was held proper when plaintiffs established that the drums were unsealed just before their contents were added to the spray solution and that no contaminants came from the water or other sources (*Eaton Fruit Co.* v. *California Spray Chemical Corp.* 1968, *Burr* v. *Sherman Williams Co.* 1954). Increasingly, liability for adulterated chemicals is imposed under the Restatement of Torts 2nd Section 420A theory of strict liability for defective products [*Shields* v. *Morton Chemical Co.* (1973)].

The utility of a pest management consultant industry and the likelihood that both consultant and grower will be equally able to assess the low-probability risks of program failure suggest that there is no reason to hold the industry to a standard of strict liability. For purposes of evaluating the potential liability of pest management services, it should be recognized that strict liability does not make the party held liable the victim's insurer. It is increasingly recognized that the relevant question raised in the decision to apply strict liability to a defendant is: which party is in the best position to make a cost-benefit evaluation between accident costs and accident avoidance costs and to act on it (Calabresi and Hirschoff 1972). For this reason, assumption of risk or, in the case of warranty, disclaimer is a defense to a strict liability suit. It is possible for the defendant to show that the injured party had adequate warning about the risks and had a meaningful choice of alternatives to use or produce. It would seem that the pest management consultant and the grower would be equally able to assess the costs and benefits of a program providing there is adequate disclosure. Another reason consistent with this analysis that strict liability is imposed is to provide an incentive to manufacturers to discover defects by quality control or design changes (*Hall* v. *E. I. DuPont De Nemours and Co*. 1972). This rationale does not seem applicable to pest management either. In recent years, spreading loss on an economic base as wide as possible has been advanced as a basis for strict liability. Some courts have held defendants liable on the theory that costs of inevitable defects should be spread among all users of the product. Loss spreading is, in effect, a kind of forced insurance for all consumers. This rationale also seems inapplicable and, in fact, weighs against holding pest management advisors to a standard of strict liability.

Pest Management Districts

The success of a management program depends on uniform participation by all growers within a uniform area. Some form of collective action will be required to organize growers and to compel participation. Districts are superior to other existing techniques because legislative policy is clearly declared and all members participating in the district have notice of the extent to which growing practices will be curtailed. This contrasts with the more limited powers available to state entomologists to declare that a plant or thing is a nuisance likely to cause imminent danger to the agriculture of the state and to abate it. Such determinations may unfairly surprise farmers and are vulnerable to court challenges based on due process grounds if a prior hearing is not held. For example, the

Arizona state entomologist wished to declare the early planting of cotton a nuisance under a statute which gives the officer the power to declare plants or things a nuisance and to abate them summarily. The state Attorney General, however, advised the entomologist that the courts must determine that a nuisance, in fact, exists and that the procedure preferable to summary abatement would be to seek a court injunction requiring abatement (Op. Attorney General 59-59 construing 3 Ariz. Rev. Stat. 3-204).

Some form of pest control districts is in existence in most states and provides some precedent for the creation of management districts. However, analogies from oil and gas law appear to be more fruitful. Existing pest management schemes provide equal benefits for all who participate and do not seem to engender a great deal of opposition. The institutionalization of integrated control, on the other hand, also provides, at least in theory, equal benefits for all participating growers, but is likely to be more controversial because it is new. Thus, some form of coerced participation may be necessary. Currently districts are primarily used to create pest-free zones. In California, Cotton Control Districts may be established to prohibit cotton growing in areas where it is determined "that such prohibition is necessary for cotton pest control" (West Cal. Ag. Code 6051-6084). Districts must be formed with the consent of a percentage of the growers in the proposed restricted area. In California, 25 percent of the growers or 25 growers must petition for the formation of a district. A hearing before the state Agricultural Commissioner must be held before a district can be organized. In Texas, the Governor can proclaim a non-cotton zone and declare it a quarantine area. The shipment of cotton out of the area is prohibited (1 Vernons Texas Civil Statutes, Agriculture, Art. 71). Besides cotton pest quarantine, the primary use of districts has been to encourage chemical pesticide application by financing spraying through the levy of per acre assessments. (Arizona Rev. Stat. § 3-331. 01-10. 1973 Supp.). In Texas, pest control programs can be funded by assessments approved through a referendum among producers of specific commodities, including cotton (1 Vernons Texas Civil Statutes, Art. 55).

Voluntary pest programs exist in several states, but if compulsory participation will be necessary, the concept of compulsory pooling in oil and gas provides a useful analogy. The need to prevent the waste of oil and gas caused by the drilling of unnecessary wells has led to widespread use of compulsory unitization procedures.

Unitization allows a field to be treated as a single production unit regardless of ownership patterns. It is essential to permit secondary recovery by water flooding. An appropriate agency must make a finding that the unit operation is necessary to accomplish one or more of the purposes for which unitization may be ordered. Statutes vary, but they generally include the following common elements: (1) most require a lease holder to initiate unitization, although a few allow the state agency to initiate unitization proceedings on its own motion as well as by petition of the parties; (2) most require that the owners of a specified percentage join within a specified time period to perfect a unitization plan; and (3) some do not require the consent of the unit interests.

Unitization and pooling statutes have uniformly been found constitutional. The state may impose conservation measures on common property owners to insure that all obtain their fair share of the resource. Farmers are in a similar reciprocal relationship and the failure of one grower to join an integrated management program threatens the crops of another grower, just as the waste of oil or gas by one common supply producer threatens the right of other producers to obtain their fair share.

The above rationale is currently the basis of a proposal in Mississippi for the compulsory participation of farmers in a pest management district. Proposed legislation would provide for the following:

1. establishment of pest management districts for cotton insect suppression,
2. cotton farmer referendum,
3. authorization of the Mississippi Cooperative Extension Service as the agency to carry out the educational and organization activities necessary for proper initiation and implementation of the pest management districts,
4. authorization to organize pest management cooperatives to manage and implement the pest management program in each district,
5. authorization of the pest management cooperatives to assess and collect funds from growers to cover the cost of pest management operations in each district,
6. authorization to apply for federal funds to use for this pest management program,
7. authorization to implement the pest management program on cotton of noncooperative growers at the grower's expense,

8. authorization of fines and penalties for non-compliance,
9. establishment of a State Pest Management Advisory Committee.

The district approach would be characterized by:

(a) local petition, referendum, organization, management and assessment,
(b) no public support of the program operations, except in the case of availability of federal funds,
(c) Mississippi Cooperative Extension Service Entomologists functioning in an organizational and educational capacity: funds will be provided from the district to employ a District Extension Service Entomologist to manage and implement the authorized pest management program in the district.

EDUCATION OF FARM OPERATORS

Some future pest management tactics will be sufficiently simple and regular to be imposed on farmers through mandatory controls. Mandatory planting dates and harvesting practices, such as are now enforced in Texas, are examples of tactics which can be incorporated in the law and practiced without special training. Most new tactics, however, will be effective only under particular conditions which might develop at different times during the season. For these situations, the farmer will ultimately have the choice of recognizing and employing the concepts of integrated control or simply relying on older, less sophisticated approaches to pest management. The farmer need not become an integrated control specialist; but he must have sufficient understanding of the concepts to trust the advice of his extension agent or consultant, to carry out their recommendations, and to seek assistance when the system seems to be behaving adversely. This farmer level of understanding of the agroecosystem will hasten the adoption and improve the effectiveness of integrated control techniques.

Levels of understanding are improved through both formal education and experience. A survey of California farmers who had and had not used pest management consultants indicates that farmers with more formal education and more experience with their crop were more likely to employ a consultant than those farmers with less education and experience (Norgaard and Levinson 1974).

Regardless of whether a farmer's earlier education significantly affects his ability to comprehend agroecosystem concepts, it seems apparent that educational programs could be developed for farmers which would teach these concepts and thereby speed the adoption and effectiveness of integrated control techniques. The study by Norgaard and Levinson found three factors related to farmer use of pest management consultants: (1) the farmer's perception of the soundness of the advice given by consultants, (2) the farmer's perception of the risk of low pesticide control techniques, and (3) the frequency of contact between the farmer and the farm advisor. The extension service operates ongoing educational programs: agroecosystem concepts could be introduced to many of these programs and special classes could be developed which would be specifically designed to speed the introduction of integrated control. Given the social and environmental benefits of integrated control, these classes might justifiably be supported. In addition, this educational effort must be tailored to provide opportunities and incentives so that all groups of farmers will participate.

EDUCATION OF CONSULTANT AND EXTENSION PERSONNEL

With the increasing emphasis on the development and adoption of integrated pest management systems, a broad expertise will be demanded of extension personnel, county agents, and private pest management consultants. Whether employed by public or private agencies, the modern agricultural pest management specialist must have a sound working knowledge of the biology, ecology, and economics of the agricultural system(s) with which he is dealing. He must be aware of the uses and limitations of approved pesticides, and of the identity and biology of both major pests and secondary species and their parasites, predators, and pathogens. He must also be abreast of research developments with respect to effective integration of all the population management techniques that might be brought to bear on seasonal pest management problems. A complete integrated pest management system must include consideration of all classes of pests--insects, nematodes, plant pathogens, and weeds--that influence cotton (or other) agriculture in a given area. This approach to pest management differs quite markedly from the traditional control of pests by the application of chemical pesticides on a fixed schedule without any real assessment of the state of the pest population or the need for such chemical treatment, and with little more than passing consideration of any accompanying impact on the environment or total ecosystem within which the crop and pest occur.

Well-trained pest management personnel are needed at the present time and this demand will undoubtedly increase rapidly during the next decade. Training for pest management specialists employed by private and public agencies will be conducted principally by land grant institutions. To provide the specialized training necessary, the traditional curricula offered by these institutions will require expansion and revision. New curricula and revisions of old programs will be necessary to provide the biological, ecological, and economic expertise, and also to meet the legal requirements imposed on pesticide applicators and pest management consultants. Recent federal legislation establishes standards and licensing procedures for persons involved in almost any form of pest control. The Federal Environmental Pesticide Control Act sets standards that restrict the use of many pesticides. Under FEPCA, the Administrator of the EPA is empowered to establish standards of competence and general procedures for the certification of pesticide applicators. Several states also have passed legislation regulating entomological consultants. Of the cotton-growing states, California and Mississippi have enacted such controlling laws. California requires all private entomological consultants who prescribe the use of pesticides to be licensed, and all entomologists, private or public, must make any insecticide recommendations in writing. Mississippi law now requires that all private pest management consultants who provide field inspection services and make control recommendations to farmers hold at least a baccalaureate degree with formal training in entomology and be able to pass a state-administered licensing examination.

A number of major institutions have adopted, or are in the process of developing, pest management curricula. Some are designed as four-year programs leading to a B.S.; others are at the M.S. level, with further graduate study leading to the Ph.D. degree. The several extant curricula provide a broad range of life sciences related to agriculture and peripheral subjects, and usually require one or more growing seasons of internship in the field under the supervision of professional pest managers. The growth of these educational programs in terms of numbers of graduates and in their importance to pest management delivery system is certainly to be expected. The foreseeable need for refresher opportunities in the form of required updating of such personnel should also receive increasing emphasis, as currently primitive pest management systems grow in scope, content, and sophistication.

SOCIAL AND ECONOMIC FACTORS

Integrated pest management strategies will vary among the major cotton-producing regions due to differences in pest complexes and level of effort in the development of integrated control techniques between regions. The selection of optimal pest management strategies will depend on social and economic factors as well as biological factors. These social and economic factors include variables such as the size of cotton farms, the management sophistication of farmers, the potential to develop new sources of pest management expertise, and the strengths and weaknesses of the existing pest management delivery system in each region. Optimal new delivery mechanisms will depend on what needs to be delivered, what can be delivered, and how it will be received. Accordingly, the optimal delivery mechanism may be expected to differ between regions.

In regions where large farms, high returns, and sophisticated management predominate, new strategies can be introduced through the existing, though more highly developed, mix of private consultants and extension programs. The rate of adoption might be increased through financial incentives coupled, perhaps, with pest loss insurance. Similarly, the large size of farms and the expertise of the farmers suggest that, where group action is necessary, pest management cooperatives, rather than pest management districts, might be formed.

In areas where low levels of management expertise, small farms, and large numbers of farmers predominate, financial incentives and insurance programs to encourage the adoption of new strategies may be relatively ineffective and more costly to administer. In such areas, pest management districts, rather than cooperatives, might be superior, if new control strategies require group coordination.

This comparison of two extremely different regions indicates the general importance of the social and economic setting to the design of delivery mechanisms. To date, very little research has been undertaken on the relationships between the effectiveness of pest management institutions and social and economic variables. The following problems and hypotheses with respect to different delivery mechanisms need further study.

Collective Decision Making

Given the significance of pest management relationships between farmers, the frequency of collective decision making in the United States seems surprisingly low. There are a few examples of cooperatives and pest management districts. Mississippi is now in the process of organizing pest management districts over the entire state to facilitate boll weevil control. Collective action is more likely to take place if (1) there is a clear understanding of the pest management interrelationships by all parties, (2) all parties can benefit without an elaborate compensation mechanism, and (3) a suitable institution for decision making and enforcement already exists. The costs of establishing an organization, determining relationships among constituent farmers and optimal management rules, agreeing on how much the gainers should compensate the losers if necessary, and enforcing decisions must be less than the benefits from cooperative action for net gains to accrue. (To date, there are no examples of collective pest management action involving compensation, or even of situations where it is unclear that all parties benefit every year.)

The benefits from collective action are likely to be more obvious and easier to capture as the relationships within the agroecosystem are better determined over time. Also, collective pest management strategies may be more advantageous in the future due to economies of scale in information gathering, processing, and decision making. It may be possible to reduce the cost of collective organization, decision making, and enforcement with state or federal participation. Further research is needed to determine the potential impact of collective action on pest management and pesticide use and to determine the best way to establish pest management organizations. If pest management cooperatives or districts can reduce the load of pesticides in the environment through more efficient use, then a public subsidy program to offset organization costs may be justified.

Farm Size

Whether integrated control is delivered through consultants, extension programs, districts, cooperatives, or a combination of these mechanisms, the costs per acre for the higher levels of information and decision making decrease with increase in farm size. Field monitoring of insect populations and plant conditions must be a sampling process. The more uniform the conditions in the field, the smaller the sample proportion can be to the total. Clearly, in a large field of 320 contiguous acres managed by one farmer, the size of the sample can be considerably smaller than for 320 acres dispersed between tobacco and soybean fields and managed by 10 to 20 farmers following different practices. Similarly, decision-making costs are lower for the larger field. Frequently, only one decision has to be made for the entire field. At other times decisions may be made for 40- or 80-acre blocks. The small cotton farmer, with assistance from his consultant, extension agent, or other source of advice, must make pest management decisions every week or two, even if he has only 10 acres. Enforcement or inducement costs, if any, are also a function of the number of farms rather than acres. For these reasons integrated control will be easier to implement on larger farms than on smaller farms.

Consultants in California found that a substantial portion of their operation entails dealing with farmers rather than with insects (Norgaard and Levinson 1974). Farmers are not willing to substitute information for pesticides without understanding the information base, and the methods of decision making and prediction. In addition, the farmer wants to know how soon the consultant will return. These demands have led some consultants to establish higher rate schedules for smaller farmers who prefer more frequent contact. It appears that farms administered by consultants are larger on the average and that large farms adopt consultants (or vice versa) more readily than small farms.

REFERENCES

Burr v. *Sherman Williams Co.* (1954) 42 Cal. 2d 682, 268 p. 2d 1041.

Calabresi, T. G., and J. T. Hirschoff (1972) Toward a test of strict liability. Yale L. J. 81:1055.

Curran, G. (1959) Professional negligence: some general comments. Vanderbilt L. Rev. 535, 538.

Eaton Fruit Co. v. *California Spray Chemical Corporation* (1968) 103 Ariz. 461, 445 P. 2d 437.

Gotreaux v. *Gary* (1957) 94 So. 2d 293 Louisiana.

Hall v. *E. I. Du Pont De Nemours and Co.* (1972) 345 F. Supp. 353, 368 E.D.N.Y.

Leonard v. *Abbot* (1962) 357 S.W. 2d 778 Texas.

Leonard v. *Abbot* (1963) rev'd 366 S.W. 2d 925 Texas.

Loe v. *Lenhardt* (1961) 362 P. 2d 312 Oregon.

Mills, and Tothman (1968) Law and professional behavior: the case of Canadian chiropractors. U. of Toronto L. Rev. 18:170.

Norgaard, R. B., and A. Levinson (1974) An evaluation of integrated pest management programs for California and Arizona cotton. Prepared for CEQ-EPA.

Perry Creek Cranberry Corporation v. *Hopkins Agricultural Chemical Co.* (1966) 29 Wis. 2d 429, 139 N.E. 2d 96.

Shields v. *Morton Chemical Co.* (1973) 518 P. 2d 857 Idaho.

State v. *Ballance* (1949) 229 N.C. 764, 51 S.E. 2d 731.

White and Summers (1972) Uniform Commercial Code 289.

Young v. *Darter* (1961) 363 P. 2d 829 Oklahoma.

7

EFFECTS OF ADOPTING ALTERNATIVE PEST CONTROL PRACTICES

IMPACT ON COTTON PRODUCTION OF INTEGRATED PEST MANAGEMENT PROGRAMS

The economic and social effects of implementing integrated pest management programs will depend on the economic environment for cotton, the specific methods of control that can be adopted in the various regions, and the delivery mechanisms. The effects fall into the following categories: (1) changes in cotton production costs, cotton prices, and subsidy burdens; (2) changes in regional production advantages; (3) changes in the demand for pest management labor and material inputs; and (4) changes in the exposure of cotton farm workers to pesticide health hazards. Unfortunately, data is not available which permits any reliable comment on these effects. In a speculative manner, however, the following discussion illustrates the range of plausible effects in the interplay of various forces.

A projection of cotton production based on continuation of trends established in the past 10 years would include moderate prices with income supports and acreage restrictions. With integrated control, insect management costs in California can be reduced by about 50 percent. Extrapolating this reduction to other regions over the long run suggests that variable cotton production costs could be reduced between 5 and 10 percent. Such a cost reduction would result in an increase in the supply of cotton that farmers could produce at any given market price. Though fairly small, this increase in supply could depress domestic and international prices, unless acreage allotments were adjusted to compensate for the shift in supply or support prices were maintained. Because of reduced production costs, cotton allotments would be more valuable to the farmer as long as a price-support rather than an income-support approach was maintained. Cotton prices to domestic users should decline somewhat which would tend to offset the substitution of synthetics for cotton that has occurred over the past decade. A small decline in world cotton prices due to the adoption of integrated control in the United States could discourage foreign cotton production until integrated control strategies are implemented in those countries.

Total cotton agriculture employment related to pest management might increase or decrease slightly depending on the mix of strategies adopted. Employment in the marketing and application of chemicals would decrease roughly in proportion to the expected reduction in chemical use. The marketing and application of biological inputs might offset this decline slightly. The major offset would come from labor employed in the collection and delivery of pest management information. The employment of scouts, consultants, extension agents, and pest management personnel employed by gins, cooperatives, districts, and large farms could be significant. A slight increase in employment might result from the more sophisticated timing of farm practices which the implementation of integrated control entails. On the other hand, integrated control techniques appear to favor larger cotton farms since there seem to be economies of scale in the collection, delivery, and implementation of knowledge. Total employment as well as employment for the purpose of pest management could decrease with further increases in farm size resulting from the adoption of integrated control. Whether a net increase or decrease resulted, the total change would probably be small. The changes in the composition of employment related to pest management, however, would be significant. People currently employed in the marketing and application of chemicals may not be able to work as scouts, consultants, or extension agents due to the differences in the skills involved. If the transition to integrated control were rapid, some would be unemployed and a surplus of job opportunities would exist for others.

The fourth area of economic and social significance relates to cotton farm workers' exposure to pesticides. Economic and social losses take the form of work time (wages) foregone; medical costs borne by the worker, the employer, and the public; and discomfort--in rare instances, death--of the victim. Integrated control will use fewer pesticides. To the extent that pesticide types are similar to those used now and that substitute biological inputs involve little or no health risks, the reduction in health risk would be roughly proportional to the reduction in pesticide use. Quantitative estimates are not possible due to the paucity of reliable data on pesticides and health effects, and to uncertainties with respect to differences in reduced pesticide use and use of labor in cotton production in the various regions. Nevertheless, this is expected to be a major benefit of integrated control, and it is a benefit which will accrue largely to low-income people.

We cannot simply assume, however, that integrated control will reduce the use of pesticides leaving their mix approximately as it is today. Persistent, low toxicity insecticides such as DDT and toxaphene--including those that might be available in the future--can be much safer for cotton workers. But persistent insecticides, unfortunately, are not compatible with many short-term, fine-tuning aspects of integrated control. For example, chemical and biological inputs tend to be mutually exclusive in integrated insect control schemes because the chemical tends to wipe out the biological input or cancel its effect. Use of both inputs requires use at different times, which is increasingly difficult the more persistent the pesticide. It is persistence which allows the chemical to be effective with low toxicity and consequently low health hazard. Thus, in the case of insect pest management, the nature of integrated control implies that the mix of pesticides employed will probably be more acutely toxic than if purely chemical approaches were taken.

An alternative economic environment for cotton production might include low prices following the removal of government programs and a continuation in the historic decline of cotton prices on the world market. If the returns received by cotton farmers were to drop into the range of 20 cents to 30 cents per pound of lint, the incremental return to pest management would fall roughly proportionately and the intensity of pest management, regardless of general strategy, would decline. Under these circumstances, the social and economic gains of integrated control would be considerably less since (1) less cotton would be grown at these prices; (2) cotton production would constrict to those areas where costs are lower and where the use of pesticides and labor are already lower; and (3) the use of pesticides would decrease regardless of strategy because of the decline in price, even in these remaining areas. All of the impacts discussed under the moderate price economic environment would still occur, but at lower levels if cotton prices were lower. The relative importance of regional shifts in production due to adoption of integrated control becomes much less important under low cotton prices since these low prices, especially in the absence of government controls, would be a far more powerful inducement to regional shifts than the adoption of integrated control. Of possible importance is the consideration that, under a condition of low prices and less intensive management, integrated control might be more effective than a simple approach to the use of chemicals.

If integrated control techniques are typically divisible in an economic sense, a change in the economic environment could lead to more rapid adoption of integrated control and earlier realization of the social, economic, and environmental benefits. However, the current high and fairly fixed costs of monitoring and decision making under integrated control might counterbalance the advantages of the divisibility of integrated control techniques.

Finally, the adoption of integrated pest management programs following an increase in returns per acre due to high lint or seed prices would have other impacts on cotton production. The social and economic benefits from the adoption of integrated control under high prices would be larger for the same general reasons that they would be smaller under low prices: (1) acreage would be greater; (2) the proportion of cotton grown in the higher-cost, high-pesticide, and high-labor-use areas would be larger; and (3) in all areas the intensity of pest management would be greater because of the higher value of the cotton saved.

EFFECTS OF REGIONAL SHIFTS IN PRODUCTION

The implementation of improved pest management and other aspects of cotton technology will have considerably different effects among regions. While integrated control techniques will probably reduce production costs overall, in some regions the reductions will be small and in others integrated control might entail an increase in production costs. These real differences will cause acreage expansion in some regions and contraction in others as production relocates toward lower cost areas. Because of differences among regions in the intensity of pest problems, and the corresponding use of pesticides, these relocations will strongly affect total pesticide use. For example, one study (Dixon et al. 1973) indicates that using current pest control practices (i.e., no new integrated control programs, pesticide use could be reduced by more than 50 percent and crop production made more efficient.

A number of detailed studies have been conducted to examine the interaction between regional distribution of crop production, total pesticide usage, and crop prices. All the studies which use mathematical models calculate the most efficient regional distribution of crop production given regional yields, production costs, and various constraints on land and pesticide use.

A study by Casey and Lacewell (1973) considers only cotton and a single alternative crop in each region (grain sorghum in the West and soybeans in the East). They examine the effects of removing one or more of the following groups of pesticides from cotton production: organochlorines, organophosphates, carbamates, and anilines. Their model divides cotton production into eight large regions, most of which include several states and their estimates on increased losses and production costs by region were based on a survey of practicing crop managers.

They first calculated net returns to producers and the amounts of cotton which would be produced if the cotton acreage were fixed at its 1970 distribution and various groups of pesticides were withdrawn (Table 7-1). As more pesticides were withdrawn, total production decreased because yields were lower and acreages were fixed. Total cotton production fell from 10.3 million bales with no pesticides withdrawn to 8.7 million bales when both organochlorines and organophosphates were withdrawn (Table 7-1).

They assume that the price of cotton varies with the total amount produced with a price elasticity of -0.5. Thus, if the total cotton supply decreased by X percent, the price per pound would increase by $2X$ percent. The price of the alternative crop remains constant.

As a result of the assumed price elasticity of -0.5, the model predicts that if acreages are fixed, the net returns to the farmer will *increase* as pesticide restrictions increase. This is mainly due to the percent increase in price which is assumed to be twice the percent decrease in total production. For example, the total production drops 11 percent with the withdrawal of both organochlorines and organophosphates so the price increases 22 percent.

It should be emphasized that this large increase in price is not due to increases in production cost per pound of cotton, but instead is mainly due to a reduction in supply. However, net returns do not rise as quickly as price because costs per unit of output do rise as herbicides and insecticides are withdrawn. For example, in Table 7-1 when cotton output is reduced 6 percent and 11

TABLE 7-1 Regional Net Returns and Percentage Reduction in Cotton Output Due to Pesticide Cancellation for Fixed 1970 Acreages

Region	Actual 1970 Output (Bales)	Reduction in Cotton Output (%)				Actual Net Returns 1970	Adjusted Net Returns[a] ($1,000,000)			
		Organo-chlorines	Organo-phosphates	Organo-chlorines and Organo-phosphates	Anilines		Organo-chlorines	Organo-phosphates	Organo-chlorines and Organo-phosphates	Anilines
I. Carolinas-Tennessee	732,763	10	8	40	5	104	104	111	77	97
II. Georgia-Alabama-Florida	854,622	10	10	19	0	44	44	46	45	46
III. Mississippi-Arkansas-Louisiana	3,066,479	9	14	14	9	338	333	330	348	296
IV. Central Oklahoma, North & East Texas	1,227,158	11	23	23	14	49	58	45	59	40
V. South Texas	409,700	7	22	23	34	16	16	10	7	1
VI. Texas High Plains	1,986,944	6	10	10	9	60	61	69	79	57
VII. West Texas-New Mexico	303,190	2	4	18	15	9	13	15	11	5
VIII. Arizona-California	1,810,030	1	4	9	0	4	23	36	32	12
Total U.S.	10,290,880	6	11	14	7	623	648	663	658	553
Price increase per pound of lint (¢)		2.7 (9,667)[b]	4.8 (9,166)[b]	7.7 (8,758)[b]	3.1 (9,567)[b]					

[a] Net returns are producer net returns for cotton and the alternative crop to cotton.
[b] Total U.S. bales in thousands.

SOURCE: Casey and Lacewell 1973a.

percent, returns increase only 4.0 percent and 6.4 percent respectively.

In a second projection, Casey and Lacewell do not restrict acres to the 1970 pattern. Rather, they calculate how much cotton would be produced in each region if, in each region, growers chose between producing cotton and the alternative crop on the basis of which would earn the larger net return. The net return to cotton depends on the yield which is a function of pesticide availability and on the price which is a function of the total amount of cotton produced. The net return of the alternative crop remains constant in each region.

Column 2 of Table 7-2 indicates the optimal allocation of crop production if no pesticides are withdrawn. Column 1 gives the actual regional distribution of acreage. They calculate that a more efficient acreage distribution would shift cotton production out of Carolina, Georgia, Alabama, and Florida; reduce cotton acreage in Mississippi, Arkansas, Louisiana, Arizona, and California; and increase cotton production in Oklahoma, Texas, and New Mexico. Presumably, these shifts have not occurred because incentives from the land retirement program and the costs (social, as well as economic) of transferring production have retarded them.

The other columns in Table 7-2 show the optimal allocation of production given the withdrawal of certain groups of pesticides. A withdrawal of organochlorines or anilines encourages a shift from south Texas to Arizona and California. Under this model, because acreage adjustments are allowed, more cotton is produced than under the fixed acreage model; thus, prices are lower. The total net returns decrease as pesticide restrictions are increased.

One of the significant findings of the Casey and Lacewell report is that cotton production (even under current pesticide usage) would be much more efficient with regional shifts in production. A similar result was obtained from a different model based on 1965 data by Heady and Brokken (1968). This study was based on a linear programming model which calculated the number of acres in each region that should be planted in one of a number of crops (including cotton) to meet 1965 demands and minimize total costs (transportation as well as production costs). Results indicated that 1965 production levels (on all crops) could be met on 178 million acres, instead of the 260 million actually used in 1965, if production were distributed more efficiently (Table 7-3).

TABLE 7-2 Regional Cotton Production Distribution and Net Returns Due to Pesticide Cancellations and Free Acreage Allotment

Region	Acres[a] (x1,000)				Net Returns ($1,000,000)			
		Adjusted				Adjusted		
	1970	No Cancellations	Organochlorines	Anilines	1970	No Cancellations	Organochlorines	Anilines
I	788	0	0	0	104	142	142	142
II	900	0	0	0	44	82	82	82
III	2,603	662	662	662	338	432	429	446
IV	1,839	3,420	3,420	3,224	49	95	94	75
V	509	464	0	0	16	31	29	29
VI	2,348	4,605	4,605	2,449	60	134	104	115
VII	217	581	581	581	9	35	36	20
VIII	953	663	1,015	1,530	4	47	47	57
U.S.	10,291 (10,291)[b]	10,396 (9,229)[b]	10,284 (9,071)[b]	8,446 (8,546)[b]	623	999	963	967

[a]Acreage adjustments were estimated using the unrestricted routine of the simulation model; i.e., cotton produced in regions where per acre cotton net returns exceed that of the alternative crop to cotton and vice versa.
[b]Total U.S. bales in thousands.

SOURCE: Casey and Lacewell 1973a.

TABLE 7-3 Regional Production Distribution Calculated To Be Most Efficient

Region	Actual 1965 Crop Acreage			Heady-Brokken Acreage Allocation		
	Feed Grains[a]	Cotton	Other[b]	Feed Grains	Cotton	Other
Northeast	4.5	--	5.7	3.7	--	3.9
Lake states	16.7	--	8.3	11.8	--	7.4
Corn Belt	38.2	0.3	30.4	32.0	--	22.1
Northern Plains	25.6	--	35.0	6.8	--	29.4
Appalachian	5.3	0.9	7.7	3.0	--	2.8
Southeast	4.3	1.9	3.1	0.1	--	1.5
Delta states	1.4	3.2	8.2	0.3	[c]	4.8
Southern Plains	10.7	6.4	13.4	3.6	9.8	9.8
Mountain	5.7	0.5	13.4	0.7	0.8	12.3
Pacific	3.6	0.7	5.7	1.2	1.3	9.3
TOTAL	116.3	14.1	130.5	63.2	12.0	103.3

[a] Feed grains include corn, oats, barley, and grain sorghum.
[b] Other includes wheat, soybeans, tame hay, and wild hay. Heady and Brokken (1968) give the acreages for each of these crops separately, but since they are not important insecticide users, they are aggregated.
[c] Less than 0.05 million acres.

SOURCE: Heady and Brokken (1968); Dixon (1973).

Dixon et al. (1973) analyzed the effect this more efficient distribution would have on total insecticide use. Based on the amount of insecticide per acre used in 1965 on crop "i" in region "j," they estimated the total amount of insecticide which would be used under the more efficient Heady-Brokken distribution.

Their results indicated that insecticide use would be reduced more than 50 percent by allocating production in the fashion designated by Heady and Brokken as most efficient. Dixon et al. (1973) developed indices which measured the contributions of crop location, acreage reduction, and crop mix to reduced pesticide use. Both crop location and acreage indices were negative which indicated that in the Heady-Brokken production distribution, crops were situated in regions that used fewer insecticides and that the reduction in acreage (by 32 percent) significantly reduced insecticide use. The crop mix index was slightly positive which indicates that there was a small shift to more insecticide intensive crops.

It should be emphasized that Dixon et al. (1973) are predicting that reduced insecticide use would be a natural consequence of a more efficient distribution of production. Their study does *not* assume any compulsory reduction or restriction in insecticide use as do the Casey-Lacewell and Pimentel-Shoemaker studies.

An important aspect of the Heady-Brokken allocation of production is a shift of cotton production from East to West. These results are similar to the Casey and Lacewell results except for the calculation that production in Arizona and California (S.P.) would decrease rather than increase. Dixon (1973) calculated that the shift of cotton from East to West alone is responsible for 70 percent of the total insecticide reduction on crops considered. This is understandable since almost half of the insecticides used in U.S. agriculture are used on cotton.

None of these studies deals with the considerable influence of the land retirement program on crop distribution. Heady (1972) expanded the Heady-Brokken model to include a land retirement program, more producing regions, and new water storage facilities. Since the model was to be used for long-range planning, it was designed to describe the year 2000 with demands, costs, and yields projected to that date.

Pimentel and Shoemaker (1974) modified the Heady (1972) model to examine the effects on costs and land use patterns if all insecticides were removed from cotton and corn production. A summary of the production distributions calculated to be most efficient are presented in Table 7-4.

TABLE 7-4 Comparison of Production Cost and Land Use for Three Situations[a]

Situation	Cotton Lint Price	Cotton Acreage	Corn Grain Price	Corn Grain Acreage	Total Acreage
Situation A[b]	$0.14/1b[d]	4.43M	$0.93/bu	30.80M	1213.6
Situation B[b]	$0.23/1b[e]	5.16M	$1.38/bu	33.81M	1221.8
Situation C[c]	$0.15/1b[f]	4.71M	$0.94/bu	31.56M	1215.7

[a]Situation A, free land market with insecticides; Situation B, land treatment program with insecticides; Situation C, free land market limiting insecticides used on cotton or corn.
[b]From Heady 1972.
[c]From Pimentel and Shoemaker 1974.
[d]$65.54/480-1b bale.
[e]$111.48/480-1b bale.
[f]$72.49/480-1b bale.

TABLE 7-5 Geographical Distribution of Cotton Production Resulting from Various Restrictions on Land Use or Insecticide Use

Consuming Region	Situation A[b] (1000 acres)	Situation B[b] (1000 acres)	Situation C[c] (1000 acres)
Atlanta	1,090	652	751
Memphis	28	1,552	28
New Orleans	2,963	1,680	3,266
Cincinnati	0	1	0
Denver	0	20	52
Dallas	38	98	38
Amarillo	22	405	22
El Paso	3	378	213
Houston	0	17	0
Los Angeles	281	276	339
San Francisco	0	75	0
TOTAL	4,429	5,159	4,712

[a] See Table 7-5 for definitions of situations.
[b] From Heady 1972.
[c] From Pimentel and Shoemaker 1974.

Comparison of the results of these two studies indicates that land use restrictions are more important in causing higher costs and acreages than restrictions on insecticides. Without acreage controls (in which case, crops could be grown in regions most suited for their production), national crop production, with or without insecticides, was more efficient and approximately 50 percent less expensive than production under a land retirement program. Without acreage controls, the calculated costs of growing corn and cotton were 1 percent and 11 percent greater, respectively, and required 2.1 million acres more land than with insecticide use (Table 7-4).

The Heady (1972) study predicts an increase in cotton production in the Delta (New Orleans) area when acreages are unrestricted (Situation A *vs.* B in Table 7-5), while the earlier version of the same model (Heady and Brokken 1968) predicted a decrease in production in this area. This difference is apparently due to the fact that the demands for other crops, especially feedgrains, are predicted to have increased and, thus, forced some western production back into the Delta area. With restrictions on insecticide (Situation C of Table 7-5), even more acreage is grown in the Delta because insect losses along the Delta are lower than in some other regions.

REFERENCES

Casey, J., and R. D. Lacewell (1973) Estimated impact of withdrawing specified pesticides from cotton production. So. J. Agric. Econ. 5:153-159.

Dixon, O. (1973) Insecticide requirements in an efficient agricultural sector. Rev. Econ. Stat. 55:423-432.

Heady, E. O., and R. F. Brokken (1968) Interregional adjustment in crop and livestock production, a linear programming analysis. Tech. Bull. No. 1396. USDA-ERS.

Heady, E. O. (1972) Future water and land use: effects of selected public agricultural and irrigation policies on water demand and land use. Tech. Rep. Center Agric. Rural Dev. Iowa State University.

Pimentel, D., and C. A. Shoemaker (1974) An economic and land-use model for reducing insecticides on cotton and corn. Environ. Entom. 3:10-20.

8

RELATIVE ENVIRONMENTAL STRESS OF PRODUCING SYNTHETIC FIBERS IN PLACE OF COTTON AND OF GROWING ALTERNATIVE CROPS

Cotton production uses more insecticide than any other crop. Cotton is also one of the few crops for which a reasonable synthetic substitute can be produced without pesticides. Synthetic fibers do not yet completely substitute for the textile uses of cotton, but these fibers probably could make up for any foreseeable reduction in the availability of cotton fiber. Cotton is heavily subsidized directly, via the government subsidy program, and indirectly, by such devices as low rates for irrigation water. One obvious way to reduce the use of insecticides, therefore, is to substitute synthetic fiber for cotton. Even if reduced use of pesticides were not an explicit societal goal, there is now enough pressure to reduce subsidies to agriculture that such substitution might occur as a side effect of economic policies made for other reasons.

Consequently, it would be valuable to compare the overall environmental effects of cotton and synthetic fiber production. We will not have made an overall gain in environmental quality if the substitution results, for example, in greatly increased air pollution or excessive demands on limited petroleum resources, in exchange for reduced pesticide use. Unfortunately, a thorough analysis is beyond the resources of the committee. Furthermore, there is some doubt that the necessary data are available, even if we had the resources to analyze them. Our object, therefore, is to indicate the directions such an analysis might take.

COTTON

Figure 8-1 presents a flow sheet for cotton production. Environmental effects arise directly from the activities shown: pesticide use, erosion, eutrophication, salinization, habitat destruction from cultivation, fertilizer use, irrigation, and air pollution and occupational disease from ginning. In addition, a number of "indirect" effects occur. For example, energy is used and air pollution, water pollution and habitat destruction are produced in the obtaining and manufacturing of fertilizers.

None of these effects, however, would disappear if synthetic fiber production were to replace some cotton production, since the major result would likely be that other crops would be planted in place of cotton over much of the released acreage. On the other hand, some land probably would go out of crop production, especially marginal land. To the extent that land would go out of crop production some environmental effects would be ameliorated. Otherwise, an analysis of the relative impact of cotton and alternative crops would be needed. Since the alternatives vary from one geographic area to another, the geographic pattern of reduced cotton acreage would have to be projected. Most of the alternative crops currently use less insecticide than cotton does, but their irrigation and fertilizer requirements are variable.

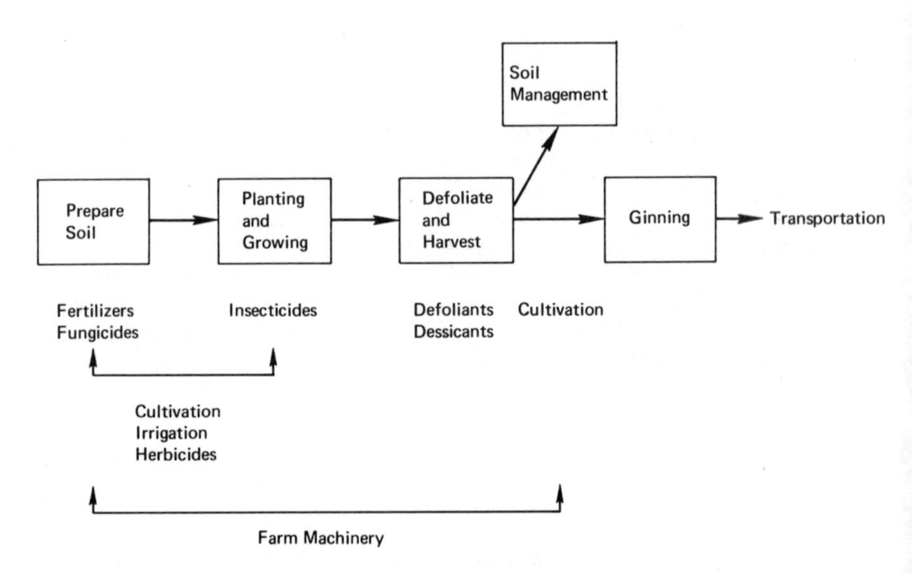

FIGURE 8-1 Flow chart for cotton.

SYNTHETIC FIBERS

Cotton can be replaced in some of its uses by fibers based either on cellulose or petroleum. The former make a relatively small market contribution which is likely to decline, so the following analysis considers only synthetic fibers, i.e., those based on petroleum products.

Synthetic fiber production occurs in two processes: first, petroleum refining that produces the petrochemical feedstocks used to produce the fibers; and second, the production of fibers from these chemical feedstocks. Petroleum refining and the processes of obtaining and shipping oil produce environmental effects such as oil spills and air pollution. Petroleum refining also produces some water pollution and solid waste.

A 1973 EPA report notes that the waste water from the synthetic fiber industry contains moderate amounts of BOD5 and COD, while nylon production in particular has waste high in organic nitrogen. Some slude is also produced.

In addition to a comparison of the environmental effects of producing cotton lint and synthetic fiber, a comparison should be made of: (1) waste production during the processing of the two types of fibers in the textile industry, (2) total energy required for their production and use, and (3) the effects of their disposal. At least the first of these comparisons should be relatively easy since the EPA has made a study of textile industry water pollution. An analysis of waste production and energy use should relate these activities to national waste production and energy use to provide some idea of the relative impact of these activities.

Finally, future technological changes may affect the comparison and should be considered. For example, it is likely that control of waste from point sources can be greatly improved, which would affect the synthetic fiber and textile effluent, while changes in pest control practices are likely to reduce insecticide use on cotton. If cotton were grown mainly for its protein value, not only would pest control practices be likely to change, but synthetic fiber then could not be a substitute product.

REFERENCE

U.S. Environmental Protection Agency. Report EPA 440/1-73/010, 1973.

/SB608.C8E591975>C1/